Springer Theses

Recognizing Outstanding Ph.D. Research

Aims and Scope

The series "Springer Theses" brings together a selection of the very best Ph.D. theses from around the world and across the physical sciences. Nominated and endorsed by two recognized specialists, each published volume has been selected for its scientific excellence and the high impact of its contents for the pertinent field of research. For greater accessibility to non-specialists, the published versions include an extended introduction, as well as a foreword by the student's supervisor explaining the special relevance of the work for the field. As a whole, the series will provide a valuable resource both for newcomers to the research fields described, and for other scientists seeking detailed background information on special questions. Finally, it provides an accredited documentation of the valuable contributions made by today's younger generation of scientists.

Theses are accepted into the series by invited nomination only and must fulfill all of the following criteria

- They must be written in good English.
- The topic should fall within the confines of Chemistry, Physics, Earth Sciences, Engineering and related interdisciplinary fields such as Materials, Nanoscience, Chemical Engineering, Complex Systems and Biophysics.
- The work reported in the thesis must represent a significant scientific advance.
- If the thesis includes previously published material, permission to reproduce this must be gained from the respective copyright holder.
- They must have been examined and passed during the 12 months prior to nomination.
- Each thesis should include a foreword by the supervisor outlining the significance of its content.
- The theses should have a clearly defined structure including an introduction accessible to scientists not expert in that particular field.

More information about this series at http://www.springer.com/series/8790

Manel Hammami

Level Doubling Network and Ripple Correlation Control MPPT Algorithm for Grid-Connected Photovoltaic Systems

Doctoral Thesis accepted by
The University of Bologna, Bologna, Italy

Springer

Author
Dr. Manel Hammami
Department of Electrical, Electronic
and Information Engineering
University of Bologna
Bologna, Italy

Supervisors
Prof. Gabriele Grandi
Department of Electrical, Electronic
and Information Engineering
University of Bologna
Bologna, Italy

Prof. Massimo Rudan
Department of Electrical, Electronic
and Information Engineering
University of Bologna
Bologna, Italy

ISSN 2190-5053 ISSN 2190-5061 (electronic)
Springer Theses
ISBN 978-3-030-10491-7 ISBN 978-3-030-10492-4 (eBook)
https://doi.org/10.1007/978-3-030-10492-4

Library of Congress Control Number: 2018965437

This Springer imprint is published by the registered company Springer Nature Switzerland AG
The registered company address is: Gewerbestrasse 11, 6330 Cham, Switzerland

Supervisors' Foreword

It is a great pleasure to introduce Dr. Manel Hammami's Ph.D. thesis work, accepted for publication within Springer Theses and awarded with a prize for an outstanding original work in 2018.

Dr. Hammami won a Ph.D. scholarship from the Erasmus Program "EU-Metalic II" under the supervision of Profs. Massimo Rudan and Gabriele Grandi. In October 2017, she completed the third and last Ph.D. year, and on April 27, 2018, she defended the Ph.D. thesis entitled *Level Doubling Network and Ripple Correlation Control MPPT Algorithm for Grid-Connected Photovoltaic Systems*, receiving the rating "Excellent" from the evaluation board. She successfully worked as part of the research group of Prof. Grandi, as witnessed by the scientific publications she coauthored, always giving a fundamental contribution to the research activities she was involved in.

From the very beginning, she expressed strong motivation, interest, and dedication to extend her knowledge and improve her skills in the field of power electronics and photovoltaics. She also demonstrated the ability to properly work with other colleagues in an international environment, always in a cooperative and friendly way. She quickly integrated into the research team, also thanks to her fluent usage of many languages (English and Italian, apart from the mother tongues Arabic and French).

She started her Ph.D. research period with a survey on maximum power point tracking (MPPT) techniques. In particular, the implementation of ripple correlation control RCC-MPPT algorithm in single-phase PV systems was examined. Further on, a hybrid RCC-MPPT was proposed to smooth out the instability introduced by fast irradiance transients and to overcome the estimation error of the voltage derivative of power (dP/dV) in the basic RCC-MPPT algorithm. RCC-MPPT algorithm was implemented together with the dq grid current controller, and both the simulation and experimental validations have been presented.

In this work, she reports the improvements of the output-power quality provided by a multilevel inverter with level doubling network (LDN). Modulation principles and harmonic analysis of output voltages are proposed and introduced in detail for both single- and three-phase LDN configurations. Then, a detailed analysis of the dc-link current and voltage ripple for H-bridge and LDN cells is carried out. In particular, the peak-to-peak dc-link current and voltage ripple amplitudes are calculated analytically over the fundamental period as a function of the modulation index for both the low-frequency and the switching ripple components. In this analysis, reference is made to a sinusoidal output current with unity power factor, representing the working condition of most of the grid-connected inverters. Maximum peak-to-peak values of both low-frequency and switching-frequency voltage ripple components are effectively adopted to design the capacitors of both H-bridge and LDN cells, basing on the desired dc-link voltage ripple requirements. The analysis is supplemented with numerical and experimental results.

The implementation of maximum power point tracking (MPPT) schemes by the ripple correlation control (RCC) algorithm, in case of multiple ripple harmonics, is also discussed. Different estimations of dP/dV are proposed in case of multiple harmonics, and a specific case with level doubling network (LDN) is considered. The steady-state and transient performances of the proposed RCC-MPPT schemes have been numerically tested and compared by MATLAB/Simulink. Results have been verified by experimental tests considering the whole multilevel PV generation system.

The work of Dr. Hammami has been the base for many scientific papers; it also produced a number of fruitful concepts, suitable for a variety of applications in the field of power electronic circuits.

Bologna, Italy
September 2018

Prof. Gabriele Grandi
Prof. Massimo Rudan

Parts of this thesis have been published in the following documents:

International Conferences

- **M. Hammami**, G. Grandi, M. Rudan, "An Improved MPPT Algorithm Based on Hybrid RCC scheme for Single-Phase PV Systems," 42nd Annual Conf. IEEE Ind. Electron. Soc., IECON 2016, October 24–27 2016, Florence, Italy.
- M. Vujacic, M. Srndovic, **M. Hammami**, G. Grandi, "Evolution of DC Voltage Ripple in Single-Phase H-bridge PWM Inverters," 42nd Annual Conf. IEEE Ind. Electron. Soc., IECON 2016, October 24–27 2016, Florence, Italy.
- M. Vujacic, **M. Hammami**, M. Srdnovic, G. Grandi, "Evolution of DC Voltage Ripple in Three-Phase PWM Voltage Source Inverters," 26th IEEE International Symposium on Industrial Electronics, ISIE 2017, June 19–21 2017, Edinburgh, UK.
- **M. Hammami**, G. Grandi, M. Rudan, "RCC-MPPT Algorithms for Single-Phase PV Systems in Case of Multiple DC Harmonics," 6th International IEEE Conference on Clean Electrical Power, ICCEP 2017, June 27–29 2017, Liguria, Italy.
- **M. Hammami**, S. Toretti, G. Grandi, "Thermal Analysis of a PV Module with Integrated Energy Storage System," 16th International Conference on Sustainable Energy Technologies, SET 2017, July 17–20 2017, Bologna, Italy.
- **M. Hammami**, M. Vujacic, G. Grandi, "Dc-Link Current and Voltage Harmonics in Three-Phase Three-Level Flying Capacitor Inverters with Sinusoidal Carrier-Based PWM," 19th International IEEE Conference on Industrial Technology, ICIT 2018, February 20–22 2018, Lyon, France.
- **M. Hammami**, M. Vujacic, A. Viatkin, G. Grandi, "Analysis of a Flexible Single-Phase Multilevel Inverter Topology for Photovoltaic Applications", 9th International Renewable Energy IEEE Congress, IREC 2018, March 20–22 2018, Hammamet, Tunisia.
- M. Vujacic, **M. Hammami**, O. Dordevic, G. Grandi, "Evaluation of Dc-Link Voltage Ripple in Five-Phase Voltage Source Inverters", 9th International IET-IEEE Conference on Power Electronics, Machines and Drives, PEMD 2018, April 17–19 2018, Liverpool, UK.

International Journals

- P. Sanjeevikumar, G. Grandi, F. Blaabjerg, P. Wheeler, P. Siano, **M. Hammami**, "A Comprehensive Analysis and Hardware Implementation of Control Strategies for High Output Voltage DC-DC Boost Power Converter," Int. Journal of Computational Intelligence Systems, IJCIS (I.F. 1.1), vol. 10, no. 1, 140–152, 2017.

- M. Vujacic, **M. Hammami**, M. Srndovic, G. Grandi, "Theoretical and Experimental Investigation of DC Voltage Switching Ripple in Single-Phase H-bridge PWM Inverters," MDPI Energies (I.F. 2.2), vol. 10, no. 8, art. 1189, 2017.
- **M. Hammami**, S. Torretti, F. Grimaccia, G. Grandi, "Thermal and Performance Analysis of a Photovoltaic Module with an Integrated Energy Storage System," MDPI Applied Science (I.F. 1.7), vol. 7, no. 11, art. 1107, 2017.
- **M. Hammami**, G. Grandi, "Single-Phase Multilevel PV Generation System with Improved Ripple Correlation Control MPPT Algorithm," MDPI Energies (I.F. 2.2), vol. 10, no. 12, art. 2037, 2017.
- M. Vujacic, **M. Hammami**, M. Srndovic, G. Grandi, "Analysis of dc-Link Voltage Switching Ripple in Three-Phase PWM Inverters," MDPI Energies (I.F. 2.2), vol. 11, no. 2, art. 471, 2018.

Acknowledgements

It is my honor to express my gratitude to all the people that came across during the Ph.D. studies at University of Bologna.

First and foremost, I would like to express my deep gratitude to Prof. Gabriele Grandi and Prof. Massimo Rudan, my research supervisors for the continuous support to my Ph.D. study and research, for their patience, motivation, and enthusiasm.

My deepest gratitude goes to Prof. Gabriele Grandi, who has unselfishly helped me with all the issues we have come across during my Ph.D. research activity. I was very lucky to have had the opportunity to work with an extraordinary person such as himself. Among many difficulties we had to solve, most appreciable was his help from academic point of view. His professional supervision and immense technical knowledge have helped my research to steer in the right direction and framed this thesis. His support and guidance have always given me strength to continue forward.

I especially thank Prof. Massimo Rudan, for his guidance during my research. He has always been accessible and willing to help me. He always encouraged me during weak moments to overcome them and to powerfully and decisively carry on.

All the merits for the results obtained during my research period should be addressed to them. I am looking forward to continuing our collaboration at the high academic and professional level in the future.

I would like to thank Prof. Alessandro Vanelli Coralli who has been with us—Ph.D. students—all the time and who has always given many technical comments on my work and evaluated it as the good one.

I also want to take a moment to thank Prof. Carlo Angelo Borghi and Prof. Ugo Reggiani for investing time and providing interesting, valuable, and wise feedbacks.

My sincere thanks also go to Prof. Claudio Rossi, Prof. Luca Zarri, Prof. Antonio Morandi, Prof. Gabriele Neretti, and Prof. Giovanni Mazzanti, for their kind support.

I wish to express my sincere appreciation for the financial support I have received during my Ph.D. studies from the Erasmus Program EU-Metalic II. Its coordinator Livia Mercatelli is a compassionate and responsible person who has been cooperating with all students at a high professional and academic level.

I wish to thank the department technicians, and in particular the electrical engineers Marco Landini, Alessio Pilati, and Matteo Marano for sharing with me their technical experience and advising me for specific technical issues that I have faced during the experimental tests.

I would like to thank Dr. Gabriele Rizzoli and Dr. Davide Pontara for helping me and advising me for specific technical issues that I have faced realizing and designing an interface board. I wish to express my sincere appreciation for the friendly support I have received during my Ph.D. studies from my colleagues Marija Vujacic, José Alberto, and Paolo Seri and for all our wonderful and inspiring meetings.

It has been a huge privilege to be with the Department of Electrical, Electronic and Information Engineering "Guglielmo Marconi," University of Bologna. Collaborating with all professors and colleagues and sharing the same willingness have been priceless for me.

Finally, I must express my very profound gratitude to my family for providing me unfailing support and continuous encouragement throughout my years of study and through the process of researching and writing this thesis. Thanks to my father, for all the interest that he has shown in my education ever since I was a child. This accomplishment would not have been possible without him. Furthermore, I would like to thank all my friends for all the care and support they provided me.

Contents

Chapter 1
General Introduction

1.1 Motivation for Research

Energy demand has increased enormously, and the consumption of fossil fuels and nuclear energy has caused many environmental problems. In fact, fossil fuels are exhaustible, polluting and rather expensive. However, the development of clean and renewable energies such as wind, biomass, geothermal and photovoltaic, are a promising solution to overcome problems of conventional energy sources. Currently, photovoltaic (PV) energy is one of the most widely used form of renewable energy and is becoming one of the main power sources [1, 2]. Nowadays, single-stage single-phase PV inverters have widely been used for their characteristics of compact configuration and modular design mainly in low-power applications (i.e., up to 5–10 kW).

It is known that maximum power point tracking (MPPT) algorithms are adopted to improve the efficiency of the PV arrays. Many methods have been proposed in the literature, such as perturb and observe (P&O) [3–6], incremental conductance (IC) [7–9], constant voltage, fuzzy logic [10, 11], etc. For these algorithms the typical problems are the identification of an appropriate perturbation step size, and the poor MPP tracking capability during sudden variations in sun irradiance. Most of these methods belong to the class of P&O algorithms, using either fixed or variable step size to improve the settling time and MPP resolution.

In single-stage single-phase systems, the instantaneous power oscillations (100 Hz) can be exploited as internal perturbation to determine the MPP of the PV field. This method, referred to in the literature as ripple correlation control (RCC), leads to good performances in tracking the MPP. RCC is generally simple, fast, and does not require any external action to perturb the PV operating point compared to the others MPPT algorithms.

The basic RCC-MPPT algorithm has been proposed in [12–14]; two high-pass filters (HPF), and two low-pass filters (LPF) are required for the implementation. In order to overcome the problem of defining the time constant of the filters, a

M. Hammami, *Level Doubling Network and Ripple Correlation Control MPPT Algorithm for Grid-Connected Photovoltaic Systems*, Springer Theses,
https://doi.org/10.1007/978-3-030-10492-4_1

modified RCC-MPPT algorithm has been introduced in [15]. Similar methods are presented in [2] in which the moving average (*MAvg*) concept has been used instead of high/low-pass filters to identify PV current and voltage oscillations (100 Hz ripple components). Furthermore, only the sign of the product of PV power and voltage ripple can be used to drive the operating point toward the MPP instead of the estimation of the voltage derivative of power (*dP/dV).*

In recent years, multilevel inverters have become more attractive for single and three-phase systems [16–21]. The most common multilevel converter topologies, presented in literature, are the neutral-point-clamped (NPC), flying capacitor (FC) and cascaded H-Bridge (CHB) converters. In both NPC and FC configurations, the number of additional components (diodes or capacitors) proportionally increases with the number of levels, leading to lower reliability, higher complexity, volume, and cost. Increasing the number of levels using the cascaded H-bridge configuration is a flexible solution, it doesn't require additional components but it needs an isolated dc power source for each H-bridge cell. Recently, asymmetric cascaded H-bridge multilevel inverter topology has gained interest from many researchers for PV system applications [22, 23]. It is based on a modular half-bridge (two switches) cascaded to a full H-bridge in order to double the output voltage levels (also called level doubling network, LDN). A proper PWM pattern provides for a self-balancing mechanism keeping the floating capacitor voltage around the half of the dc-link H-bridge voltage. The LDN configuration (Fig. 1.1) is becoming popular, due to its simple, modular and reliable structure and can be considered as a retrofit which can be added to existing H-bridge configurations in order to double the output voltage levels.

The concept of LDN has the capability of self-balancing the dc capacitor voltage during positive and negative fundamental cycles without any closed-loop control/

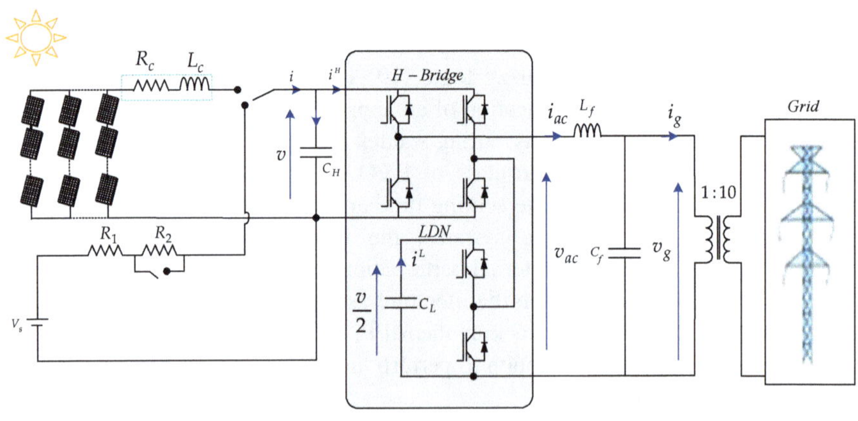

Fig. 1.1 Block diagram of single-phase LDN multilevel inverter (H-bridge and LDN)

algorithm, and it does not consume or supply any active power (reactive inverter). It significantly improves the power quality, reducing the cost and size of the harmonic filter.

Despite numerous ripple correlation control RCC-MPPT algorithms for single-stage single phase grid-connected PV systems have been developed, no detailed analysis of RCC algorithms in case of multilevel inverters has been reported since the multilevel configuration introduces additional PV voltage and current low-frequency harmonics, perturbing the basic implementation of the RCC scheme, leading to malfunctioning. This dissertation is devoted towards the analysis of voltage and current harmonic components used in RCC schemes in order to estimate dP/dV and to drive the PV operating point to the MPP in case of multiple harmonics introduced in a more precise and faster manner.

1.2 Research Objectives

The main objective of this dissertation is to study, analyze and improve the stability and the dynamic performance of Ripple Correlation Control (RCC) Maximum Power Point Tracking (MPPT) algorithms in case of multiple harmonics in case of grid-connected multilevel inverters. Verifications have been carried out by numerical simulation software and a real experimental setup. In detail, the objectives are focused towards:

(1) Improving the capability of the RCC-MPPT algorithm under sudden and large variation of sun irradiance in case of single-phase H-bridge grid-connected PV systems.
(2) Improvements of output power quality by multilevel inverter with level doubling network (LDN). Modulation principles and harmonic analysis of output voltages have been proposed and introduced in details for both single and three-phase LDN configurations.
(3) Analysis of dc-link voltage ripple in single-phase LDN configuration. Calculation of peak-to-peak input current and voltage amplitudes for both low and switching frequency components have been given. Guidelines for designing the dc-link capacitors have been proposed based on these calculations. Numerical and experimental verification of the results have been obtained.
(4) The analysis of dc-link voltage ripple has been extended also to three-phase multilevel inverters, specifically three-phase cascaded H-bridge configuration (up-to 9-level) and three-phase LDN configurations (up-to 17-level). The peak-to-peak dc-link current and voltage ripple diagrams are presented and discussed. Simulation verification of the results have been obtained.
(5) The implementation of maximum power point tracking (MPPT) schemes by the ripple correlation control (RCC) algorithm is discussed in case of multiple

ripple harmonics, as in the case of multilevel inverters. Different possible *dP/dV* estimations have been proposed in case of multiple harmonics, and a specific case with level doubling network (LDN) has been considered. The steady-state and transient performances of the proposed RCC-MPPT schemes have been numerically tested and compared by MATLAB/Simulink. Results have been verified by experimental tests considering the whole multilevel PV generation system.

By achieving the objectives listed above, a significant new knowledge has been produced. This has been evidenced by the already published research papers that have resulted from the thesis, which can be found within cited references.

Chapters 2–7 contain the original results from the research and therefore represent the main contribution of this work.

1.3 Outlines and Original Contributions of the Dissertation

Apart the Introduction (this chapter), the thesis is composed of different chapters according to the following organization:

Chapter 2

Basic RCC-MPPT Scheme in Single-Phase H-Bridge Grid-Connected PV Systems

This chapter provides a survey on maximum power point tracking (MPPT) techniques and proceeds with the basic implementation of ripple correlation control RCC-MPPT algorithm in single-phase PV systems. Further on, a hybrid RCC-MPPT has been proposed to smooth out the instability introduced by fast irradiance transients and to overcome the estimation error of the voltage derivative of the power (*dP/dV*) in the basic RCC-MPPT algorithm. RCC-MPPT algorithm has been implemented together with the *dq* current controller and both the simulation and experimental validations have been presented.

Chapter 3

Improvement of Output Power Quality by Level Doubling Network (LDN)

This chapter reports the improvements of output power quality by multilevel inverter with level doubling network (LDN). Modulation principles and harmonic analysis of output voltages have been proposed and introduced in details for both single and three-phase LDN configurations.

Chapter 4

Analysis of dc-link Current and Voltage Ripple: Single-Phase Configuration

This chapter provides a detailed analysis of the dc-link current and voltage ripple for H-bridge and LDN cells. In particular, the peak-to-peak dc-link current and voltage ripple amplitudes are analytically calculated over the fundamental period as a function of the modulation index for both the low-frequency and the switching ripple components. Reference is made to sinusoidal output current with unity power factor, representing the working condition of the most of grid-connected inverters. Maximum peak-to-peak values of both low-frequency and switching frequency voltage ripple components are effectively adopted to design the capacitors of both H-bridge and LDN cells basing on the desired dc-link voltage ripple requirements. The analysis is accompanied with the numerical and experimental results.

Chapter 5

Analysis of dc-link Current and Voltage Ripple: Three-Phase Configuration

This chapter gives a complete analysis of dc-link voltage ripple for three-phase configurations. First presented is the dc-link current and voltage ripple for three-phase H-bridge configuration in case of both sinusoidal and centered PWMs. Further on, the analysis is extended to three-phase LDN configuration. In particular, a detailed analysis of the dc-link current and voltage ripple for H-bridge and LDN cells is provided. The peak-to-peak dc-link current and voltage ripple amplitudes are analytically calculated over the fundamental period as a function of the modulation and in case of sinusoidal output current with unity power factor, representing the working condition of the most of grid-connected inverters. Based on the DC voltage requirements, simple and effective guidelines for designing DC-link capacitors are obtained for both H-bridge and LDN cells. Results have been numerically tested by MATLAB/Simulink.

Chapter 6

Improvement of RCC-MPPT in Case of Multiple dc-link Voltage Harmonics

The implementation of maximum power point tracking (MPPT) schemes by the ripple correlation control (RCC) algorithm, in case of multiple ripple harmonics, is discussed in this chapter. Different possible estimations of the voltage derivative of the power (*dP*/*dV*) have been proposed in case of multiple harmonics, and a specific case with level doubling network (LDN) has been considered. The steady-state and transient performances of the proposed RCC-MPPT schemes have been numerically tested and compared by MATLAB/Simulink. Results have been verified by experimental tests considering the whole multilevel PV generation system.

Chapter 7

Conclusion and Perspectives

The main conclusions of this dissertation based on simulation and experimental results are presented. Finally, chapter also provides few recommendations for future research work.

References

1. S.B. Kjaer, J.K. Pedersen, F. Blaabjerg, A review of single-phase grid-connected inverters for photovoltaic modules. IEEE Trans. Ind. Appl. **41**(5), 1292–1306 (2005)
2. M. Hammami, G. Grandi, M. Rudan, An improved MPPT algorithm based on hybrid RCC scheme for single-phase PV systems, in *IECON 2016—42nd Annual Conference of the IEEE Industrial Electronics Society* (2016), pp. 3024–3029
3. N. Femia, G. Petrone, G. Spagnuolo, M. Vitelli, Optimization of perturb and observe maximum power point tracking method. IEEE Trans. Power Electron. **20**(4), 963–973 (2005)
4. M. Khaled, H. Ali, M. Abd-El Sattar, A.A. Elbaset, Implementation of a modified perturb and observe maximum power point tracking algorithm for photovoltaic system using an embedded microcontroller. IET Renew. Power Gener. **10**(4), 551–560 (2016)
5. R.M. Linus, P. Damodharan, Maximum power point tracking method using a modified perturb and observe algorithm for grid connected wind energy conversion systems. IET Renew. Power Gener. **9**, 682–689 (2015)
6. L. Piegari, R. Rizzo, I. Spina, P. Tricoli, Optimized adaptive perturb and observe maximum power point tracking control for photovoltaic generation. Energies **8**(5), 3418–3436 (2015)
7. E. Kandemir, N.S. Cetin, S. Borekci, A comprehensive overview of maximum power extraction methods for PV systems. Renew. Sustain. Energy Rev. **78**, 93–112 (2017)
8. C. Li, Y. Chen, D. Zhou, J. Liu, J. Zeng, A high-performance adaptive incremental conductance MPPT algorithm for photovoltaic systems. Energies **9**(4), 288 (2016)
9. D. Sera, L. Mathe, T. Kerekes, S.V. Spataru, R. Teodorescu, On the perturb-and-observe and incremental conductance MPPT methods for PV systems. IEEE J. Photovoltaics **3**(3), 1070–1078 (2013)
10. S.E. Babaa, M. Armstrong, V. Pickert, T. Esram, P.L. Chapman, Overview of maximum power point tracking control methods for PV systems. IEEE Trans. Energy Convers. **22**(08), 59–72 (2014)
11. T. Esram, P.L. Chapman, Comparison of photovoltaic array maximum power point tracking techniques. IEEE Trans. Energy Convers. **22**(2), 439–449 (2007)
12. J.W. Kimball, P.T. Krein, Discrete-time ripple correlation control for maximum power point tracking. IEEE Trans. Power Electron. **23**(5), 2353–2362 (2008)
13. D. Casadei, G. Grandi, C. Rossi, Single-phase single-stage photovoltaic generation system based on a ripple correlation control maximum power point tracking. IEEE Trans. Energy Convers. **21**(2), 562–568 (2006)
14. R. Khanna, Q. Zhang, W.E. Stanchina, G.F. Reed, Z.H. Mao, Maximum power point tracking using model reference adaptive control. IEEE Trans. Power Electron. **29**(3), 1490–1499 (2014)
15. C. Boonmee, Y. Kumsuwan, Control of single-phase cascaded H-bridge multilevel inverter with modified MPPT for grid-connected photovoltaic systems, in *IECON 2013—39th Annual Conference of the IEEE Industrial Electronics Society* (2013), pp. 566–571

16. E. Babaei, S. Laali, Z. Bayat, A single-phase cascaded multilevel inverter based on a new basic unit with reduced number of power switches. IEEE Trans. Ind. Electron. **62**(2), 922–929 (2015)
17. G. Buticchi, E. Lorenzani, G. Franceschini, A five-level single-phase grid-connected converter for renewable distributed systems. IEEE Trans. Ind. Electron. **60**(3), 906–918 (2013)
18. G. Buticchi, D. Barater, E. Lorenzani, C. Concari, G. Franceschini, A nine-level grid-connected converter topology for single-phase transformerless PV systems. IEEE Trans. Ind. Electron. **61**(8), 3951–3960 (2014)
19. X. Yuan, H. Stemmler, I. Barbi, S. Member, Self-balancing of the clamping-capacitor-voltages in the multilevel capacitor-clamping-inverter under sub-harmonic PWM modulation. IEEE Trans. Power Electron. **16**(2), 256–263 (2001)
20. N.A. Rahim, J. Selvaraj, Multistring five-level inverter with novel PWM control scheme for PV application. IEEE Trans. Ind. Electron. **57**(6), 2111–2123 (2010)
21. S. Debnath, J. Qin, B. Bahrani, M. Saeedifard, P. Barbosa, Operation, control, and applications of the modular multilevel converter: a review. IEEE Trans. Power Electron. **30** (1), 37–53 (2015)
22. S.K. Chattopadhyay, C. Chakraborty, A new asymmetric multilevel inverter topology suitable for solar PV applications with varying irradiance. IEEE Trans. Sustain. Energy **3029**(99), 1 (2017)
23. S.K. Chattopadhyay, C. Chakraborty, A new multilevel inverter topology with self-balancing level doubling network. IEEE Trans. Ind. Electron. **61**(9), 4622–4631 (2014)

Chapter 2
Basic RCC-MPPT Scheme in Single-Phase H-Bridge Grid-Connected PV Systems

2.1 Introduction

A maximum power point tracking (MPPT) control is necessary in photovoltaic (PV) generation system in order to extract maximum power from the PV arrays. Several MPPT techniques have been proposed and discussed in the literature. The most famous MPPT algorithms are based on perturb and observe (P&O) [1–4], incremental conductance (IC) [1, 5, 6], open-Circuit Voltage [1], fuzzy logic [1, 7] etc. The primary challenges of the MPPT methods are the improvement of the settling time, stability, sensors required, simplicity, cost, MPP tracking capability and other aspects. Among many papers presented in the literature, much focus has been on perturb and observe (P&O) [2–4, 8]. The identification of an appropriate perturbation step size, and the poor MPP tracking capability during sudden variations in sun irradiance limit the use of the basic P&O and IC techniques. For this reason, alternative solutions have been proposed. A modified P&O algorithms have been proposed [2, 4, 9] and a tradeoff between accuracy and speed in selecting the step size has been considered to improve the efficiency of these methods belong to the class of P&O algorithms.

In single-phase systems, the instantaneous power oscillations (100 Hz) can be exploited as internal perturbation to determine the MPP of the PV field. This method, referred to in the literature as ripple correlation control (RCC), leads to good performances in tracking the MPP. RCC-MPPT algorithm was proposed in [10]. In [11, 12], two high pass filters (HPF), and two low pass filters (LPF) are required for the implementation of the RCC-MPPT algorithm. Moreover, only the sign of the product of power and voltage ripple is used as a control signal to drive the operating point toward the MPP instead of the estimation of dP/dV.

RCC is generally simple, fast, and does not require any external perturbation compared to the other MPPT algorithms. However, the literature does not provide

M. Hammami, *Level Doubling Network and Ripple Correlation Control MPPT Algorithm for Grid-Connected Photovoltaic Systems*, Springer Theses,
https://doi.org/10.1007/978-3-030-10492-4_2

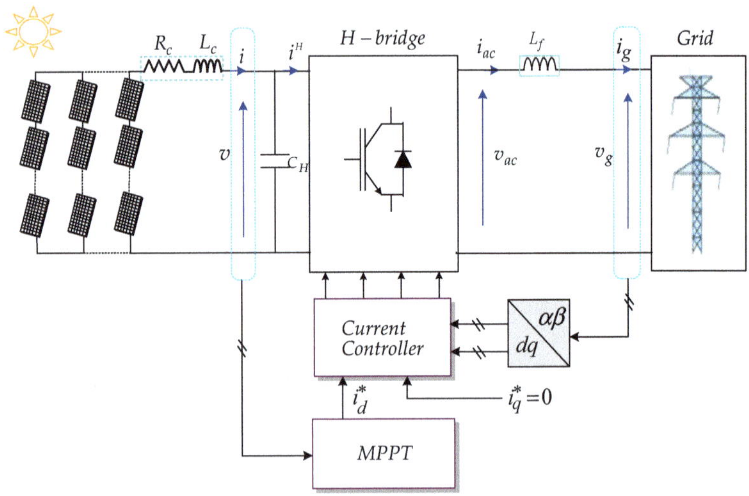

Fig. 2.1 Block diagram of the single-phase PV generation system

analysis to improve the capability of the algorithm to track the MPP under sudden and large variation of sun irradiance.

A modified RCC-MPPT method able to smooth out the instability introduced by fast irradiance transients is proposed [13]. In particular, a "transient detector" is introduced, able to switch the control mode between the standard RCC mode, in steady-state operation, and the constant-voltage mode, during transients. In this way, the proposed control scheme combines the fast response capabilities of RCC with the stable behavior of constant-voltage controller during large sun irradiance transients. A fast *dq* current controller for regulating the power injected into the grid is considered to optimize the dynamic performance [14]. The block diagram of the whole single-phase PV generation system is shown in Fig. 2.1.

2.2 Basic Implementation of RCC-MPPT Algorithm

With reference to the single-phase grid-connected PV generation system of Fig. 2.1, neglecting all the losses and assuming a unity power factor, the instantaneous ac power is described as:

$$p_{ac} = v_{ac} i_{ac} = V_{ac} \cos \omega t \, I_{ac} \cos \omega t = \frac{1}{2} V_{ac} I_{ac} (1 + \cos 2\omega t) \tag{2.1}$$

where V_{ac} and I_{ac} are peak values of inverter voltage and current, and $\omega = 2\pi f$ is the grid angular frequency ($f = 50$ Hz).

Neglecting the inverter losses, the switching ripple, and supposing the inverter input voltage almost constant equal to ($v = V$), the input/output power balance is considered in (2.2):

$$V\bar{i}^{H} = mV\cos\omega t I_{ac}\cos\omega t \tag{2.2}$$

where m is the inverter modulation index ($m = V_{ac}/V$) and $\bar{i}^{H}$ is the input averaged current over the switching period (T_{sw}), representing the low frequency input current harmonics, expressed as:

$$\bar{i}^{H} = I^{H} + \tilde{i}^{H} \tag{2.3}$$

being I^{H} the dc (average) component and $\tilde{i}^{H}$ the alternating double-fundamental frequency component, expressed and readily obtained by (2.1) and (2.2):

$$I^{H} = \frac{mI_{ac}}{2} \tag{2.4}$$

$$\tilde{i}^{H} = \frac{mI_{ac}}{2}\cos 2\omega t = I_{2}^{H}\cos 2\omega t \tag{2.5}$$

The grid power (2.1) reflects in the same dc power, introducing PV voltage and PV current oscillations at twice of the grid frequency (100 Hz). In case of single-stage PV systems, where the inverter is directly connected to the PV field (PV panel including the cable), the instantaneous PV voltage and current have three relevant components: dc (average working point Q), alternating double-fundamental frequency, and switching frequency components, according to:

$$\begin{cases} v = V + \tilde{v} + \Delta v \cong V + \tilde{v} \\ i = I + \tilde{i} + \Delta i \cong I + \tilde{i} \end{cases} \tag{2.6}$$

being V and I the PV voltage and current (average) components, $\tilde{v}$ and $\tilde{i}$ the alternating double fundamental frequency voltage and current components (i.e., 100 Hz), and Δv and Δi the switching frequency voltage and current components, respectively. The switching frequency component Δv is strongly smoothed by the dc-link capacitor, and it is usually negligible for switching frequencies starting from 5 to 10 kHz. It is assumed here $\Delta v \cong 0$ as well as $\Delta i \cong 0$.

Accordingly, the amplitude of the alternating component (100 Hz) of the PV voltage, V_2, can be calculated on the basis of the amplitude of the input (dc) current oscillation (2.5) and the whole dc-link impedance Z_2, as

$$V_{2} = Z_{2}I_{2}^{H} = Z_{2}\frac{mI_{ac}}{2} \tag{2.7}$$

being Z_2 the parallel impedance between the reactance of the dc-link capacitor C_H, and R_{PV} the equivalent resistance of PV field, calculated at twice of the grid frequency (2*f*):

$$Z_2 = \frac{R_{PV}}{\sqrt{1 + 4\omega^2 R_{PV}^2 C_H^2}} \tag{2.8}$$

Note that the equivalent PV resistance R_{PV} depends on the working point Q on the $I(V)$ characteristic of the PV field, and $R_{PV} \cong V_{MPP}/I_{MPP}$ in the vicinity of the MPP.

Considering realistic parameter values for PV fields and dc-link capacitors, the assumption $R_{PV} \gg 1/(2\omega C_H)$ is generally satisfied, and Eq. (2.8) becomes:

$$Z_2 = \frac{1}{2\omega C_H} \tag{2.9}$$

The amplitude of the PV current oscillation (double fundamental frequency) I_2 can be calculated based on the amplitude of the PV voltage oscillation (2.7) and the equivalent PV resistance R_{PV} as:

$$I_2 = \frac{V_2}{R_{PV}} \cong \frac{I_{MPP}}{V_{MPP}} \frac{1}{2\omega C_H} \frac{mI_{ac}}{2} \tag{2.10}$$

The dc-link voltage capacitance C_H can be designed, by (2.7), (2.9) and (2.10), on the basis of the desired PV voltage and current oscillation amplitudes, usually set in the order of few % of the MPP values. The higher the oscillation amplitude, the lower the PV efficiency, being working point moved around the MPP. The lower the oscillation amplitude, the lower the accuracy in determining the MPP, comparing to voltage and current transducers noise and resolution.

The voltage derivative of the power dP/dV for the considered working point Q can be expressed on the basis of the voltage derivative of the current dI/dV as

$$\left.\frac{dP}{dV}\right|_Q = I + \left.\frac{dI}{dV}\right|_Q V \tag{2.11}$$

The dI/dV defines the relation between current and voltage oscillation as:

$$\tilde{i} = \left.\frac{dI}{dV}\right|_Q \tilde{v} \tag{2.12}$$

The $I(V)$ characteristic has a negative slope, so current and voltage oscillations are always in phase opposition, and in case of a single oscillation harmonic (100 Hz), Eq. (2.12) can be rewritten just considering the oscillations amplitudes:

$$\left.\frac{dI}{dV}\right|_Q \cong -\frac{I_2}{V_2} \tag{2.13}$$

The most popular RCC-MPPT implementation consists in multiplying the voltage oscillation by the current oscillation, and integrating over half of grid period *T*/2 (i.e., calculating the 100 Hz moving average, *MAvg*) as:

$$\left.\frac{dI}{dV}\right|_Q = \frac{\int_{t-T/2}^{t} \tilde{i}\tilde{v}d}{\int_{t-T/2}^{t} \tilde{v}^2 dt} \tag{2.14}$$

It should be noted that, introducing the hypothesis of single voltage and current oscillation harmonics (2nd harmonic), and simplifying (2.14) leads back to (2.13) as follows:

$$\left.\frac{dI}{dV}\right|_Q = -\frac{I_2 V_2}{V_2^2} = -\frac{I_2}{V_2} \tag{2.15}$$

The block diagram representing the estimation of *dP*/*dV* by previous calculations is summarized in Fig. 2.2.

In general, the estimation of *dP*/*dV* by (2.11) is fast and precise in steady-state or "near" steady-state working conditions. Indeed, it suffers the drawback of fluctuations and inaccuracy during fast transient of sun irradiance. As a consequence, MPPT algorithms using this estimation of *dP*/*dV* as a voltage error signal in finding the MPP result in overshoots and long settling times in case of fast sun irradiance changes. This occurrence is presented in more details by simulation results given in the following.

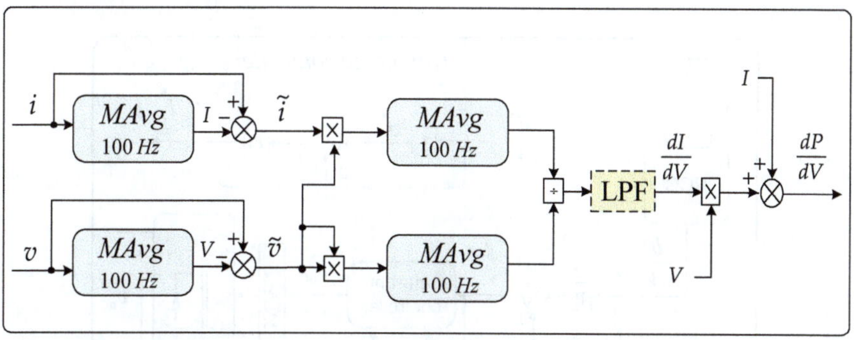

Fig. 2.2 Block diagram of the basic RCC algorithm to estimate *dP*/*dV* using a *MAvg* over *T*/2 (100 Hz)

2.3 The Proposed Regulator System

2.3.1 *Overall Control Scheme*

Figure 2.1 shows the configuration of the considered single-phase grid-connected PV system. It consists of a photovoltaic field connected to the grid by a H-bridge inverter. This structure avoids the use of intermediate dc-dc converters, increasing reliability and efficiency of the conversion system.

Figure 2.3 shows the block diagram of the whole control scheme. The RCC algorithm is used to estimate the voltage derivative of the power, *dP*/*dV*, driving the working point toward the MPP and determining the reference grid current I_g^*. In particular, I_g^* can be calculated by using either a feed-forward based on a power balance, Fig. 2.3a, or by using a voltage controller, Fig. 2.3b. A *dq* current controller regulates the PWM inverter in order to inject sinusoidal grid current with unity power factor. In particular, the inverter voltage equation at the grid side can be written as follows:

$$v_{ac} = R_f i_g + L_f \frac{di_g}{dt} + v_g \tag{2.16}$$

where v_{ac} is inverter output voltage, v_g and i_g are grid voltage and grid current, respectively; R_f and L_f are resistance and inductance of the ac-link inductor, respectively.

By applying Park's transform, (2.16) can be expressed as:

$$\begin{pmatrix} v_{acd} \\ v_{acq} \end{pmatrix} = \begin{pmatrix} R & -\omega L_f \\ \omega L_f & R \end{pmatrix} \begin{pmatrix} i_d \\ i_q \end{pmatrix} + \begin{pmatrix} L_f & 0 \\ 0 & L_f \end{pmatrix} \begin{pmatrix} \frac{di_d}{dt} \\ \frac{di_q}{dt} \end{pmatrix} + \begin{pmatrix} v_d \\ v_q \end{pmatrix} \tag{2.17}$$

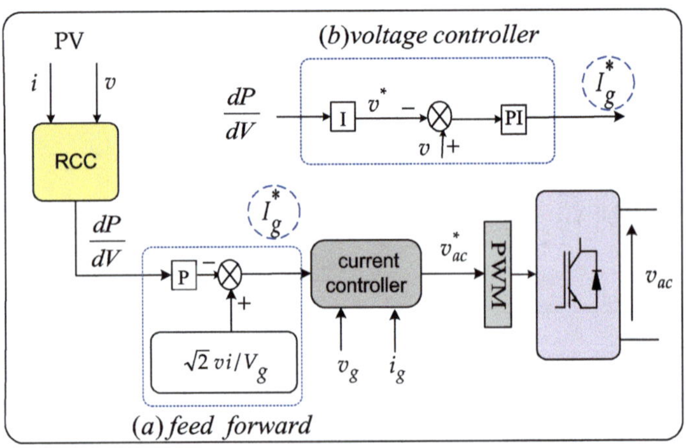

Fig. 2.3 Block diagram of the whole control scheme

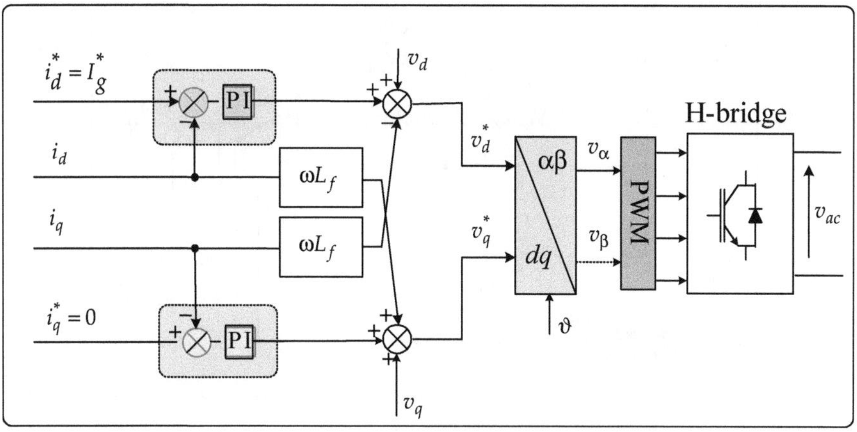

Fig. 2.4 *dq* current controller for single-phase H-bridge inverter [14]

The block diagram of the *dq* current controller applied to the considered grid-connected inverter is depicted in Fig. 2.4.

In this control strategy, grid voltage and current are transformed from the αβ reference frame to the *dq* reference frame (Fig. 2.1). A phase-locked loop (PLL) algorithm is used to determine the phase angle ϑ of the grid voltage, even in the presence of noise or higher order harmonics.

In particular, direct and quadrature currents, i_d and i_q, are used to control active and reactive powers, respectively. The reference direct current is set to I_g^*, determined by the MPPT algorithm (Fig. 2.3), whereas the reference quadrature current is set to zero in order to maintain unity power factor.

2.3.2 The Proposed Hybrid RCC-MPPT

In order to overcome the estimation error of *dP*/*dV* during fast sun irradiance transients in basic RCC-MPPT algorithms, a modified control scheme is proposed, according to Fig. 2.5. The reference grid current I_g^* is calculated by using a feed forward (Fig. 2.3a). The MPPT algorithm has two separate control modes:

- the RCC mode, active during steady-state or "near" steady state conditions ($h = 0$);
- the constant-voltage regulation mode, during large sun irradiance transients ($h = 1$).

Basically, it is a hybrid algorithm alternating between these two control modes to track the MPP for any steady-state or transient operating conditions, specifically determined by a "transient detector". In particular, the RCC mode consists in the

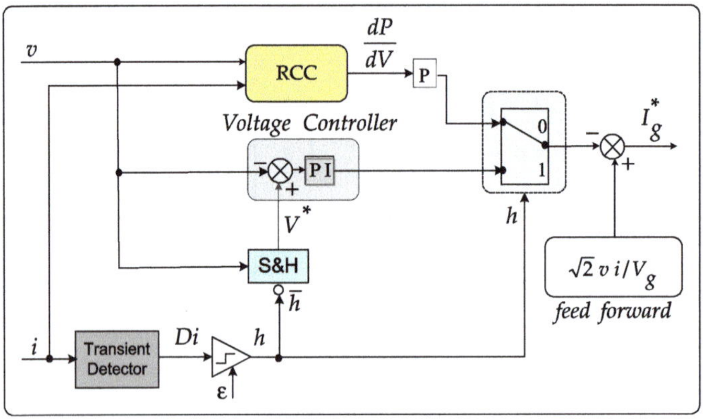

Fig. 2.5 Block diagram of the proposed hybrid RCC-MPPT algorithm

basic RCC implementation described in the Sect. 2.2, as represented in the upper part of Fig. 2.5, being active when the detector does not sense a transient ($h = 0$).

The constant-voltage mode regulator implements a dc (PV) voltage regulator, active when a transient is detected ($h = 1$), having as the voltage reference signal the last measured voltage before the transient detection. This function is implemented by a sample and hold (S&H) block driven by the opposite logic of the signal h.

There are different algorithms for detecting transients. The attention has been focused on the PV current, which is an unsmoothed variable with high dynamic.

In particular, an estimation of the PV current derivative (Di) is implemented by calculating the current variation in a given time interval (T_ε) and normalizing by the PV short circuit current (I_{sc}). The time interval should be a multiple of the 100 Hz oscillation period (i.e., 10 ms) to make the detector immune from these steady-state current oscillations. The transient sensitivity of the detector can be adjusted by setting the threshold ε for determining the detector output h as

$$h = \begin{cases} 0 & \text{if} \quad \mathrm{D}i < \varepsilon \quad \textit{steady} - \textit{state} \\ 1 & \text{if} \quad \mathrm{D}i \geq \varepsilon \quad \textit{transient} \end{cases} \tag{2.18}$$

being

$$Di = \frac{|i(t) - i(t - T_\varepsilon)|}{I_{sc}} \tag{2.19}$$

In principle, the dc (PV) voltage kept constant by the voltage reference provided by the S&H during huge sun irradiance transient ($h = 1$), since the MPP voltage doesn't change appreciably. As the sun irradiance is over ($h = 0$), the RCC-based regulator regularly tracks the MPP, the dP/dV estimation being stable and precise.

2.4 Simulation Results

The single-phase single-stage grid-connected photovoltaic system shown in Fig. 2.1 has been numerically implemented by MATLAB-Simulink on the basis of the parameters shown in Table 2.1. A string of series/parallel connected PV modules has been adopted as PV source, according to the data given in Table 2.2 (STC). In particular, PV modules are SP-305 type (96 cells, monocrystalline).

In order to evaluate the different performances between the proposed hybrid RCC scheme and the basic RCC-MPPT algorithm, the following scenarios have been considered, comparing the simulation results:

1. Ramp variation of sun irradiance (soft transient).
2. Step decreases of sun irradiance (huge transient).
3. Step increases of sun irradiance (huge transient).

For all these cases, the quantity *Di* of the transient detector defined by (2.19) has been calculated. The threshold has been set to $\varepsilon = 0.1$, and the detector is disabled in the case of basic RCC algorithm.

A. **Simulations of the basic RCC-MPPT algorithm**

For the first scenario, the initial value of irradiance is 1000 W/m^2, decreasing linearly to 500 W/m^2 in a period of 200 ms. The cell temperature has been considered constant (25 °C) during simulations. It can be seen in Fig. 2.6 that the transient is almost smoothed and the resulting estimation of dP/dV is acceptable (Fig. 2.6b). Note that *Di* always remains below the threshold $\varepsilon = 0.1$ (dashed line in Fig. 2.6c), this meaning that (2.19) did not get out the limit of the detector threshold. Also, it is clear that the PV current and the PV voltage can follow the MPP under this kind of smooth irradiance change.

Figure 2.7 shows the same transient as Fig. 2.6 in the typical *P*(*V*) diagram. The operating point moves from the first MPP, before the irradiance transient, to the new MPP, after the transient. Once the new steady-state is reached, only small oscillations persist around the new MPP.

Simulation results for step increase and decrease of sun irradiance are presented in Fig. 2.8 (left) and (right), respectively. A huge transient can be observed in both cases, causing an extremely wrong estimation of dP/dV (Fig. 2.8b), introducing over-shoots in dc link voltage (Fig. 2.8e) and large distortions in the grid current (Fig. 2.8f).

Table 2.1 Simulated system parameters

Parameter	Symbol	Value
Grid voltage (rms)	V_g	230 V
Ac-link inductor	L_f, R_f	10 mH, 0.1 Ω
PI regulator	K_P, K_I	20, 10
Dc-link capacitor	C_H	10 mF
Switching frequency	f_{sw}	10 kHz

Table 2.2 PV array specifications (STC)

Parameter	Symbol	Value
Open circuit voltage	V_{OC}	64.2 V
Short circuit current	I_{SC}	5.96 A
Maximum photovoltaic voltage	V_{MPP}	54.7 V
Maximum photovoltaic current	I_{MPP}	5.58 A
Number of series-connected modules per string	N_S	9
Number of parallel strings	N_P	3

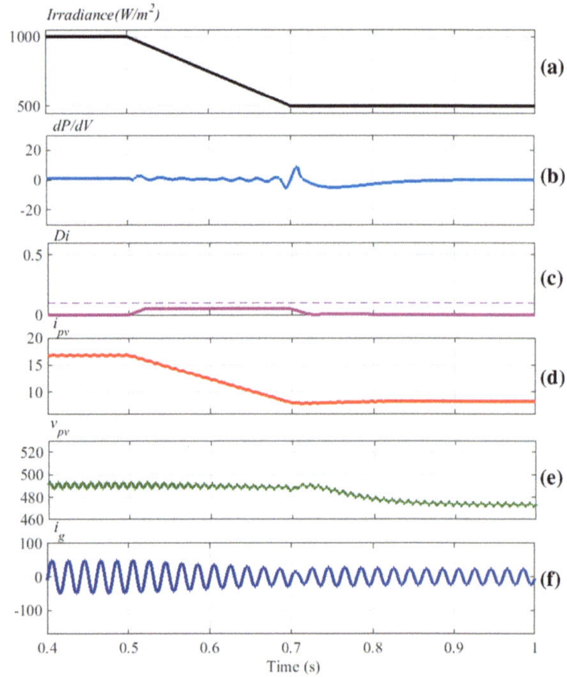

Fig. 2.6 Irradiance ramp decrease: **a** irradiance, **b** power derivative, **c** detector signal (now disabled), **d** photovoltaic current, **e** photovoltaic voltage, and **f** grid current, in case of basic RCC-MPPT algorithm

In these two cases, *Di* get out the threshold ε (dashed line in Fig. 2.8c), and the detector (now disabled) could easily recognize the transients.

At the end of the transient introduced by the step irradiance changes, i.e., after about 20–30 ms, the estimation of *dP*/*dV* becomes correct again, and the operating point is correctly driven toward the MPP. Also for these transients, the path of the operating points are displayed on the typical *P*(*V*) diagrams in Fig. 2.9.

A possible remedy to overcome the problems introduced by the wrong estimation of *dP*/*dV* is the application of a low-pass filter (LPF) to smooth the misleading oscillation of *dP*/*dV*, as shown in Fig. 2.2 (dashed rectangle). However, this solution introduces a delay in the voltage control loop, slowing down the MPP tracking speed.

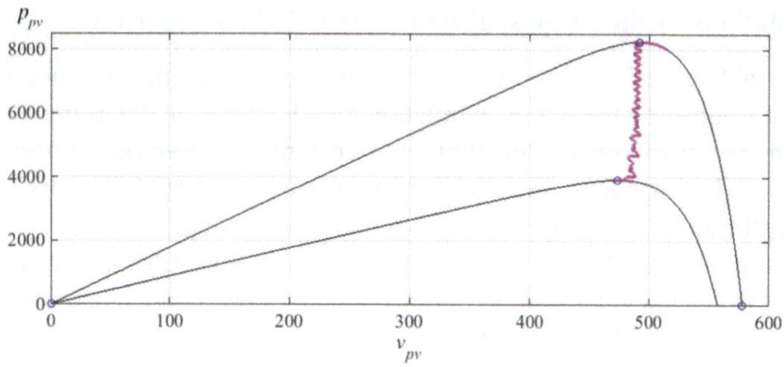

Fig. 2.7 Effects of a 50% irradiance transient of PV power versus PV voltage in case of ramp decreases in irradiance

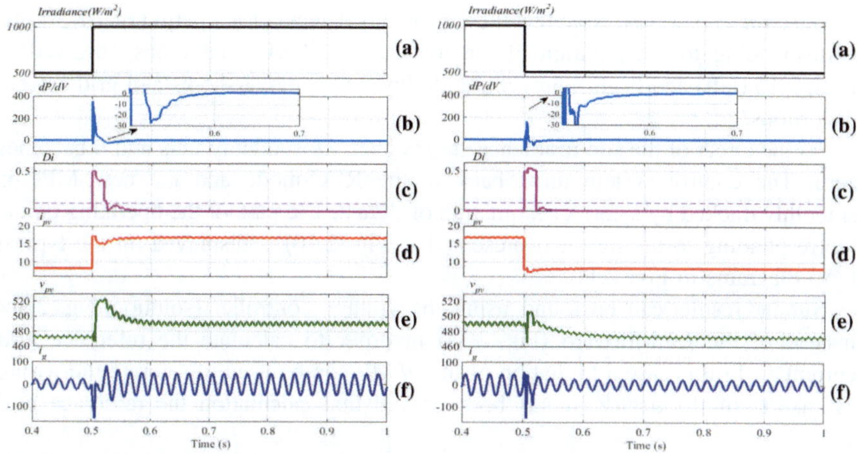

Fig. 2.8 Irradiance step increase: **a** irradiance, **b** power derivative, **c** detector signal (now disabled), **d** photovoltaic current, **e** photovoltaic voltage, and **f** grid current, in case of basic RCC-MPPT algorithm

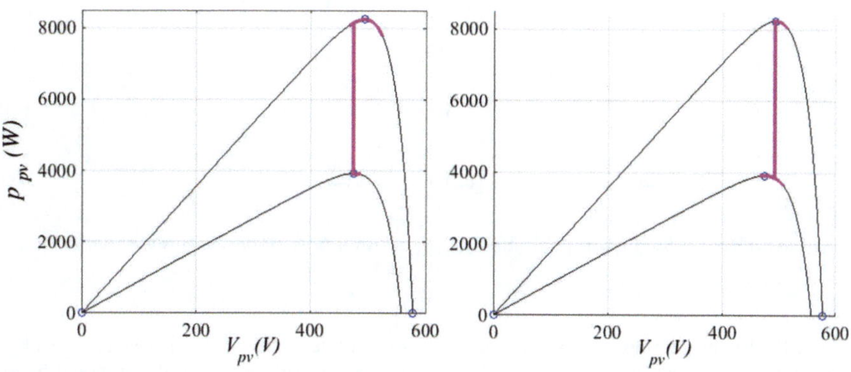

Fig. 2.9 Effects of a 50% irradiance transient of PV power versus PV voltage: step increase of irradiance (left), step decrease of irradiance (right)

B. **Simulations of the proposed hybrid RCC-MPPT algorithm**

As stated in Sect. 2.3.2, the proposed solution to overcome overshoots and distortions introduced by the wrong estimation *dP*/*dV* consist in determining when this estimation is trustworthy, selecting a constant-voltage-mode instead of the RCC-mode when a huge sun irradiance transient is detected. The RCC tracking mode is restored as the transient is recognized to be over.

The simulation results corresponding to the same sun irradiance step presented of Fig. 2.8 (left) and (right) are given in Fig. 2.10 (left) and (right), with reference to the proposed hybrid RCC scheme (Fig. 2.3).

Now the transient detector is enabled, and the control system swaps from the RCC-mode to the constant-voltage-mode as the quantity *Di* overcomes the threshold $\varepsilon = 0.1$, i.e., *h* becomes unity. The reference voltage is kept constant to the last sensed value (before the transient, when *h* was zero) by the S&H, directly driven by $\bar{h} = NOT(h)$. Note that the transient is detected for a period of 20–30 ms, corresponding to the duration of the misleading *dP*/*dV* oscillations observed in Figs. 2.10 (left) and (right), preventing dc-link voltage overshoots and grid current distortions.

As the effect of the sun irradiance step is over, *Di* suddenly falls and *h* becomes zero. The control system turns back to the RCC-mode and the new MPP is smoothly tracked by a correct estimation of *dP*/*dV*. The path of the operating points (corresponding to the transient presented in Fig. 2.10) is displayed on the typical P(V) diagrams in Fig. 2.11.

Similar results can be found using the voltage controller (solution Fig. 2.3b) instead of the feed-forward (Fig. 2.3a) in order to calculate the reference grid current I_g^*. In this case a limitation of the *dP*/*dV* can be used to saturate the wrong estimation of the *dP*/*dV*. It has been proved that calculation the reference grid

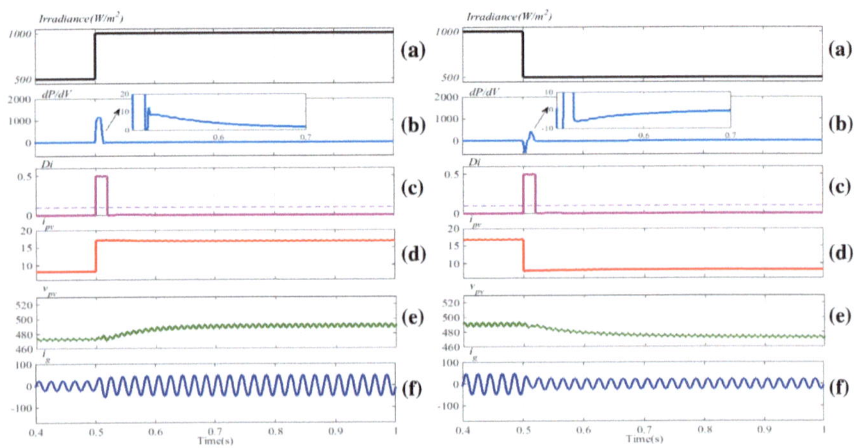

Fig. 2.10 Irradiance step increase: **a** irradiance, **b** power derivative, **c** detector signal (enabled), **d** photovoltaic current, **e** photovoltaic voltage, and **f** grid current, in case of the proposed hybrid RCC-MPPT algorithm

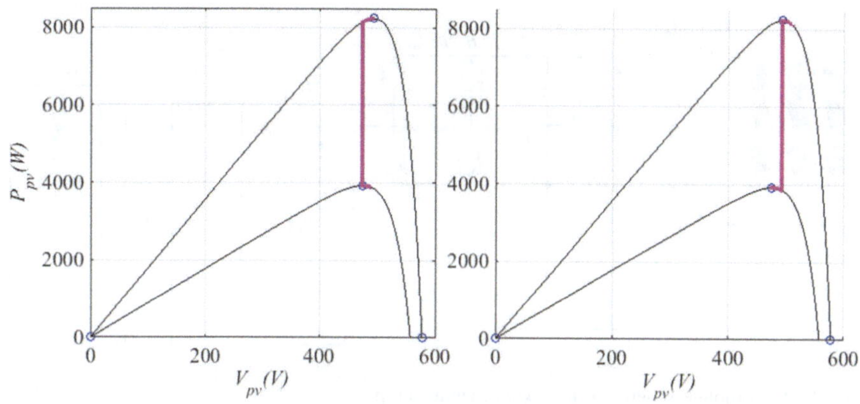

Fig. 2.11 Effects of a 50% irradiance transient of PV power versus PV voltage: step increase of irradiance (left), step decrease of irradiance (right)

current using the voltage controller is a more robust and stable solution for this reason the experimental results will be obtained by implementing the voltage controller (Fig. 2.3, scheme b).

2.5 Experimental Results

In order to verify the expected behavior of the control scheme by a simple integral voltage controller (Fig. 2.3b), preliminary experimental results are presented with reference to the basic H-bridge configuration (Fig. 2.12) and RCC-MPPT algorithm.

The employed experimental setup is shown in Fig. 2.13. It includes a DSP board (TMS320F28377D) programmed by Code Composer Studio (CCS), also employed to change all the parameters of the regulators in the real time. The single-phase H-bridge inverter has been implemented by a power IGBTs module IPM PS22A76 (1200 V, 25 A), driven by the DSP board via intermediate optical links. The inverter output is connected to the grid trough LC filter and isolation transformer with a turn ratio of 1:10. Further details about the experimental setup are given in Appendix A.

Additionally, two current transducers model LA 55-P (LEM® company) were used to measure PV current and grid current while PV voltage and the grid voltage were measured using two voltage transducers model LV 25-P (LEM® company).

Considering an operating point around the MPP, a PV module can be linearized with voltage source with a series resistance $R_{PV} = V_{MPP}/I_{MPP}$, according to the Thevenin equivalence. In order to change the operating point, preliminary experimental tests have been carried out considering a step change in the series resistances as shown in Fig. 2.12. In order to change the operating point, the transient has been

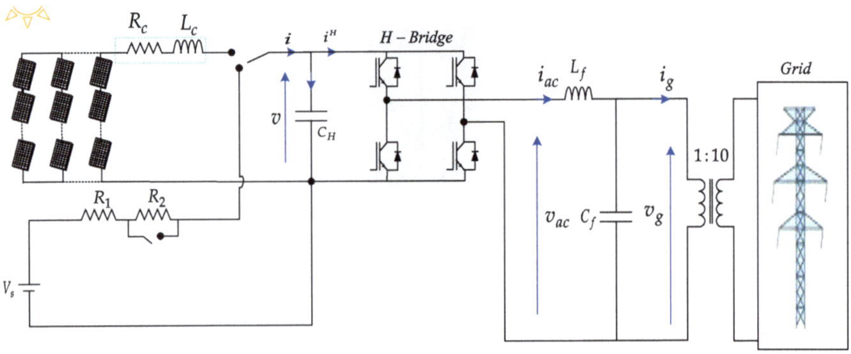

Fig. 2.12 Complete scheme of the experimental setup

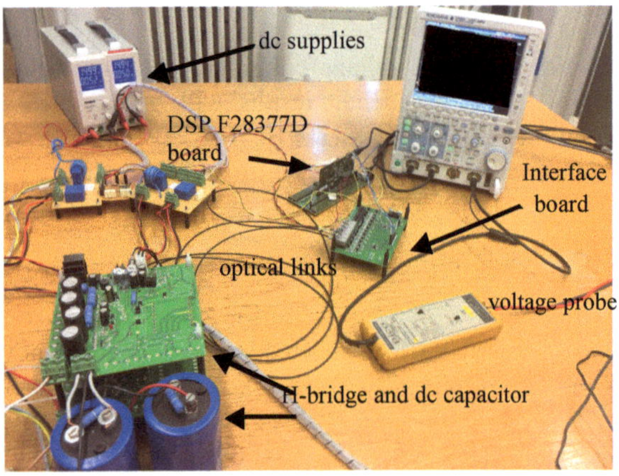

Fig. 2.13 View of the preliminary experimental setup

done by short-circuiting one of the two series resistances (R_1 and R_2). The system parameters are given in Table 2.3.

Figure 2.14 shows the steady-state behavior of the system including a dc voltage source V_s = 80 V and a series resistance $R = R_1 + R_2$.

As expected, the MPP voltage is equal to the half of the dc voltage source (40 V in this case) also it is clear that the grid current and voltage are in phase which prove the effectiveness of the current controller as well as the RCC-MPPT algorithm. The single-phase H-bridge system produces inherent current and voltage oscillations at twice of fundamental frequency (100 Hz) as known and as shown in Fig. 2.15. Both voltage and current oscillation amplitudes can easily be calculated based on Eqs. (2.7) and (2.10), respectively.

Experimental results for step increase and decrease of series resistances are presented in Fig. 2.16. In this case of transient, the operating point changes from one to another keeping the MPP voltage always around the half of the dc voltage

Table 2.3 System parameters

Parameter	Symbol	Value
Equivalent dc voltage supply	V_s	80 V
Dc source resistances	R_1, R_2	40, 20 Ω
Resistance of cable connection	R_c	0.5 Ω
Inductance of cable connection	L_c	39 μH
Dc-link H-bridge capacitance	C_H	1 mF
AC filter capacitance	C_f	25 μF
Filter resistance and inductance	R_f, L_f	0.77 Ω, 4.5 mH
Switching frequency	f_{sw}	5 kHz
Fundamental frequency	f	50 Hz

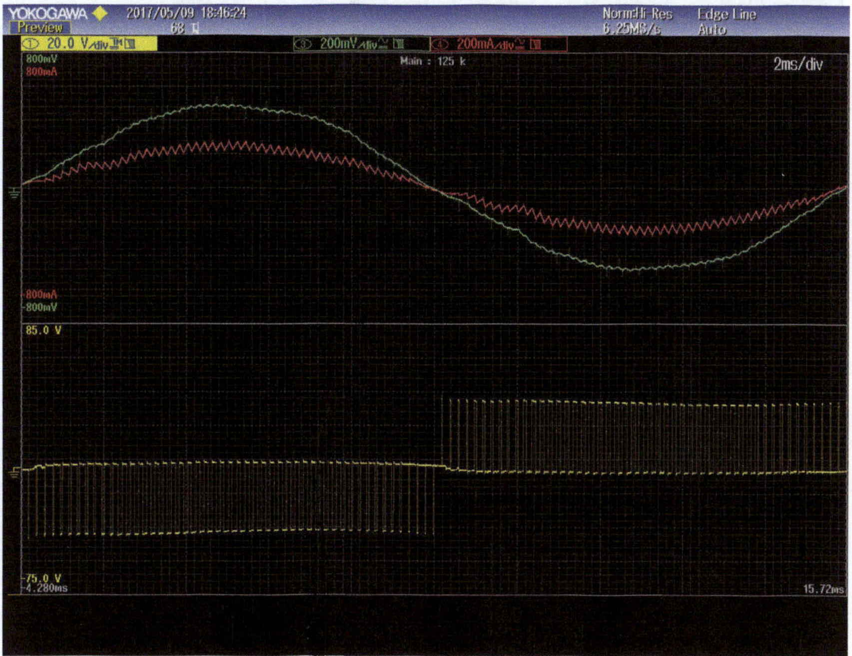

Fig. 2.14 Grid voltage and current at maximum power point (top traces) and the inverter output voltage (bottom trace)

source V_s as expected. It can be noted that the estimation of the *dP*/*dV* (blue trace) is correct and stable in steady state. However, as generally known for RCC-MPPT algorithm, it is oscillating during the step resistance transient, introducing dc-link voltage overshoots, acceptable grid current fluctuations (Fig. 2.16). The scale coefficients of the measurement quantities and the scope signals are given in Table 2.4.

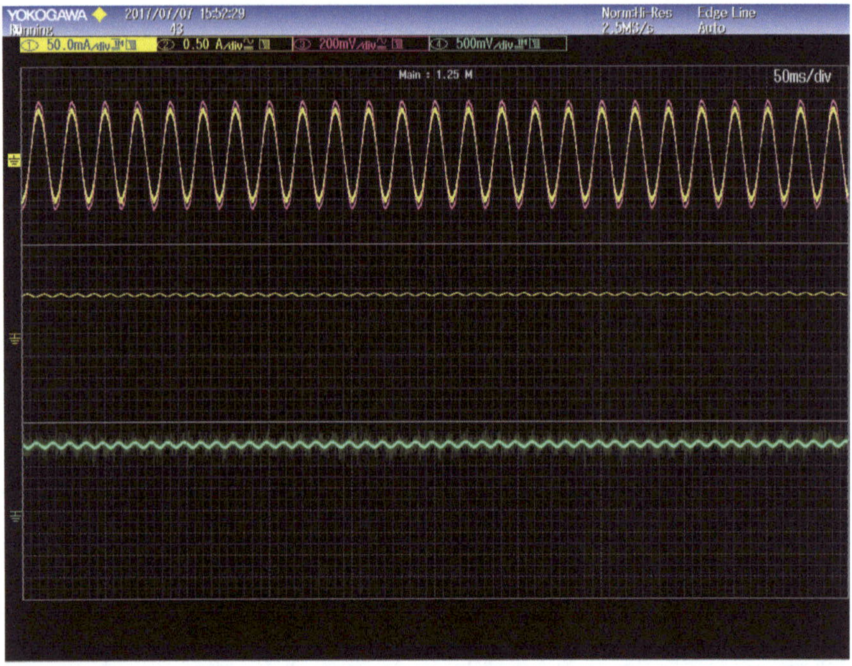

Fig. 2.15 Grid voltage and current at maximum power point (top traces), dc (PV) current (middle trace) and dc (PV) voltage (bottom trace)

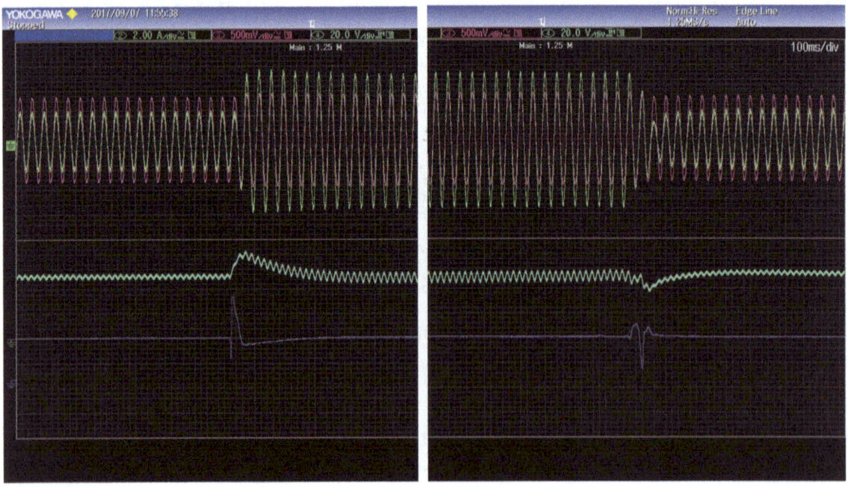

Fig. 2.16 Step increase (left) and decrease (right) in resistance. Top traces: grid current (green) and voltage (pink). Bottom traces: dc-link voltage (green) and estimation of *dP*/*dV* (blue)

Table 2.4 Scale coefficients

Transducer gains	Value
Grid current	11.79 A/3.3 V
Grid voltage	120 V/3.3 V
dc current	2.75 A/3.3 V
dc-link voltage	112 V/3.3 V

The results presented in both steady-state and transient conditions confirm the comprehensive performance of the RCC-MPPT algorithm considering the voltage controlled (Fig. 2.3, scheme b).

2.6 Discussion

As known, one of the critical aspects of the RCC-MPPT is its behavior during fast irradiance changes, since it works properly only in steady-state or nearly steady-state operating conditions, misestimating dP/dV in transient conditions. The fastest controller includes the calculation of the reference grid current by the estimated dP/dV with a feedforward compensation, according to Fig. 2.3, scheme (a). This regulation scheme suffers fast irradiance transients introducing huge oscillations in the operating point. A possible solution to overcome this drawback has been proposed and discussed, introducing a hybrid controller (Fig. 2.5) with a transient detector. Despite this solution has been found fast and effective in simulation tests, for real applications it has been found more practical to introduce a pure voltage controller, generating the reference PV voltage by a simple integral controller, even without any feedforward, according to the scheme (b) in Fig. 2.3. In this way, the integral function is able to filter itself the misestimated peaks of dP/dV, introducing lower MPPT dynamics but with more stable performances. For this reason, in the following experimental implementations (including the proposed multilevel grid inverter) the integral voltage controller has been adopted.

References

1. S.E. Babaa, M. Armstrong, V. Pickert, T. Esram, P.L. Chapman, Overview of maximum power point tracking control methods for PV Systems. IEEE Trans. Energy Convers. **22**(08), 59–72 (2014)
2. N. Femia, G. Petrone, G. Spagnuolo, M. Vitelli, Optimization of perturb and observe maximum power point tracking method. IEEE Trans. Power Electron. **20**(4), 963–973 (2005)
3. D. Sera, L. Mathe, T. Kerekes, S.V. Spataru, R. Teodorescu, On the perturb-and-observe and incremental conductance MPPT methods for PV systems. IEEE J. Photovoltaics **3**(3), 1070–1078 (2013)

4. M. Khaled, H. Ali, M. Abd-El Sattar, A.A. Elbaset, Implementation of a modified perturb and observe maximum power point tracking algorithm for photovoltaic system using an embedded microcontroller. IET Renew. Power Gener. **10**(4), 551–560 (2016)
5. C. Li, Y. Chen, D. Zhou, J. Liu, J. Zeng, A high-performance adaptive incremental conductance MPPT algorithm for photovoltaic systems. Energies **9**(4), 288 (2016)
6. M.A.M. Ramli, S. Twaha, K. Ishaque, Y.A. Al-Turki, A review on maximum power point tracking for photovoltaic systems with and without shading conditions. Renew. Sustain. Energy Rev. **67**, 144–159 (2017). (January)
7. T. Esram, P.L. Chapman, Comparison of photovoltaic array maximum power point tracking techniques. IEEE Trans. Energy Convers. **22**(2), 439–449 (2007)
8. R.M. Linus, P. Damodharan, Maximum power point tracking method using a modified perturb and observe algorithm for grid connected wind energy conversion systems. IET Renew. Power Gener. **9**, 682–689 (2015)
9. L. Piegari, R. Rizzo, I. Spina, P. Tricoli, Optimized adaptive perturb and observe maximum power point tracking control for photovoltaic generation. Energies **8**(5), 3418–3436 (2015)
10. T. Esram, J.W. Kimball, P.T. Krein, P.L. Chapman, P. Midya, Dynamic maximum power point tracking of photovoltaic arrays using ripple correlation control. IEEE Trans. Power Electron. **21**(5), 1282–1290 (2006)
11. D. Casadei, G. Grandi, C. Rossi, Single-phase single-stage photovoltaic generation system based on a ripple correlation control maximum power point tracking. IEEE Trans. Energy Convers. **21**(2), 562–568 (2006)
12. C. Boonmee, Y. Kumsuwan, Control of single-phase cascaded H-bridge multilevel inverter with modified MPPT for grid-connected photovoltaic systems, in *IECON 2013—39th Annual Conference of the IEEE Industrial Electronics Society* (2013), pp. 566–571
13. M. Hammami, G. Grandi, M. Rudan, An improved MPPT algorithm based on hybrid RCC scheme for single-phase PV systems, in *IECON 2016—42nd Annual Conference of the IEEE Industrial Electronics Society* (2016), pp. 3024–3029
14. M. Ebrahimi, S.A. Khajehoddin, M. Karimi Ghartemani, Fast and robust single-phase DQ current controller for smart inverter applications. IEEE Trans. Power Electron. **31**(5), 3968–3976 (2015)

Authored Papers

15. M. Hammami, G. Grandi, M. Rudan, An improved MPPT algorithm based on hybrid RCC scheme for single-phase PV systems, in *42nd Annual Conference of the IEEE Industrial Electronics Society, IECON 2016*, Florence, Italy, 24–27 Oct 2016
16. M. Vujacic, M. Srndovic, M. Hammami, G. Grandi, Evaluation of DC voltage ripple in single-phase H-Bridge PWM inverters, in *42nd Annual Conference of the IEEE Industrial Electronics Society, IECON 2016*, Florence, Italy, 24–27 Oct 2016

Chapter 3
Improvement of Output Power Quality by Level Doubling Network (LDN)

3.1 Introduction

In recent years, multilevel inverters have become more attractive for both single and three-phase systems, due to many advantages that they have over conventional H-bridge pulse width-modulated (PWM) inverters. They offer improved output waveforms [1–3], smaller filter size, lower total harmonic distortion (THD), higher output voltages [4] and others [5, 6].

The most common multilevel converter topologies, presented in the literature, are neutral-point-clamped (NPC), flying capacitor (FC) and Cascaded H-Bridge (CHB) converters [7–12].

In both NPC and FC configurations, the number of additional components (diodes or capacitors) drastically increases with the number of levels, leading to lower reliability, higher complexity, volume, and cost. Increasing the number of levels using the cascaded H-bridge configuration is a more flexible solution; it doesn't require additional components but it needs an isolated dc power source for each H-bridge unit.

In order to reach a compromise between the number of output voltage levels and the number of components in single-phase inverters, several hybrid and asymmetric topologies of multilevel inverters have been proposed in the literature [13–17].

In [16], a novel PWM modulation strategy has been proposed for the FC asymmetric H-bridge (FCaHB) where six power switches and one capacitor are needed as shown in Fig. 3.1 (left). A design of the flying capacitor has been presented taking into account only the dc-link voltage ripple at the switching frequency.

A NPC asymmetric H-bridge (NPCaHB) topology has been proposed in [13], but this solution needs an additional capacitor and two additional diodes comparing to FCaHB topology as shown in Fig. 3.1 (right).

M. Hammami, *Level Doubling Network and Ripple Correlation Control MPPT Algorithm for Grid-Connected Photovoltaic Systems*, Springer Theses,
https://doi.org/10.1007/978-3-030-10492-4_3

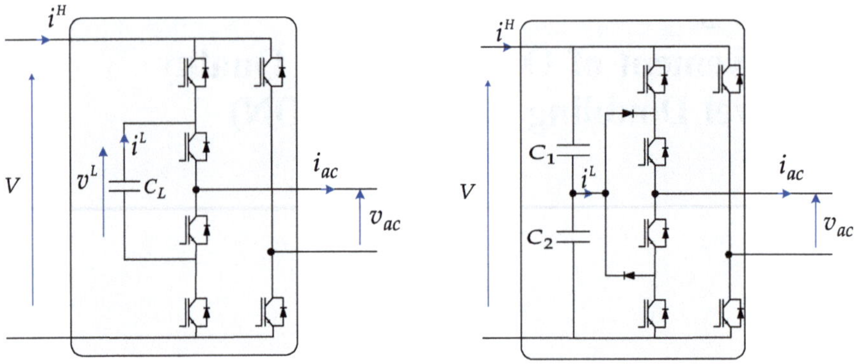

Fig. 3.1 Basic circuit schemes of the single-phase FCaHB multilevel inverter (left) and the NPCaHB equivalent counterpart (right)

Another interesting asymmetric multilevel inverter topology is the cascaded asymmetric H-bridge (CaHB), which is based on a modular half-bridge (two switches) cascaded to a full H-bridge in order to double the output voltage level (also called level doubling network, LDN). This solution is considered in this thesis and is equivalent to the FCaHB, but with a different arrangement of the half-bridge. It is becoming popular due to its simple, modular and reliable structure and it can be considered as a retrofit which can be added in series to existing H-bridge configurations in order to double the output voltage levels.

As well as single-phase ones, three-phase multilevel inverters are commonly used in many applications such as variable speed ac drives, uninterruptable power supplies (UPS), grid-connected and stand-alone renewable generation systems, etc. Different PWM control techniques have been investigated and proposed in order to reduce the output voltage and current distortion and improve other performances [18–20]. Amongst all of them, carrier based PWM (CB-PWM) and space vector PWM (SV-PWM) techniques are the most employed ones due to well-defined harmonic spectrum and fixed switching frequency. In most applications, the CB-PWM technique is preferably used due to reduced computational time and easy implementation [21].

With reference to three-phase three-level multilevel inverters, FC, NPC and H-bridge configurations are shown in Figs. 3.2 and 3.3.

The three-phase LDN topology has been presented in [5] and the concept of self balancing capability of the capacitor was analyzed. In [5, 14], only nearest voltage level control (staircase modulation) has been developed to support the LDN operation. However, the PWM modulation strategy for single and three-phase LDN inverters has not been reported in the literature yet.

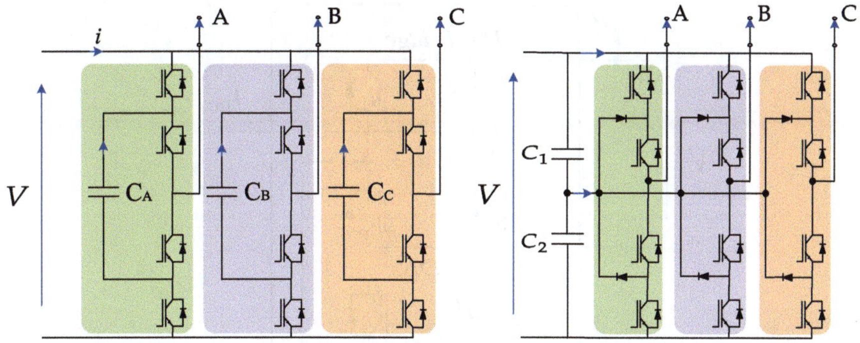

Fig. 3.2 Three-level inverter, FC configuration (left) and the NPC configuration (right)

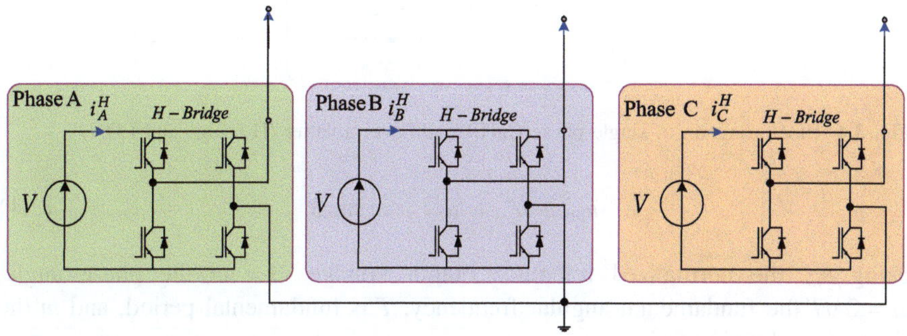

Fig. 3.3 Three-level H-bridge configuration

3.2 System Configuration and Modulation Principles

3.2.1 System Configuration

A. **Single-phase**

Figure 3.4 presents a multilevel inverter connected to a sinusoidal ac output current source (i_{ac}) with unity power factor, which is the basic assumption to represent most of the grid-connected applications. In particular, the multilevel inverter is realized by cascading the full H-bridge (supplied by a constant dc source voltage V) to an additional half-bridge (LDN) supplied by a floating capacitor to smooth the dc-link voltage oscillations.

A proper PWM pattern has been adopted, as described in the next sub-section, providing for a self-balancing mechanism. In particular, the steady-state LDN average dc-link voltage is automatically kept around the half of the dc-link voltage of the H-bridge, i.e. $v^L \approx V/2$, as proved in [5] for symmetric LDN operations.

The instantaneous output voltage v_{ac}, normalized by V and averaged over the switching period ($T_{sw} = 1/f_{sw}$), is determined within a linear modulation range as:

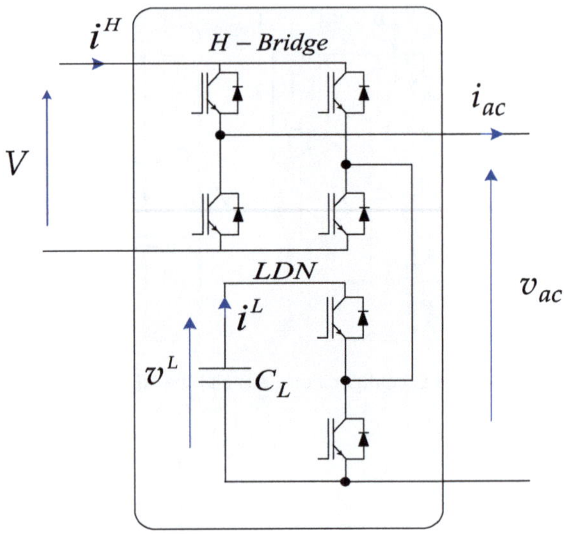

Fig. 3.4 Block diagram of single-phase CaHB multilevel inverter (H-bridge and LDN)

$$\bar{u}_{ac} \cong u^*_{ac} = u_{ac} = m \sin \vartheta \tag{3.1}$$

being u_{ac} the normalized reference output voltage, $\vartheta = \omega t$ the phase angle, $\omega = 2\pi/T$ the fundamental angular frequency, T is fundamental period, and m the inverter modulation index.

B. **Three-phase**

The single-phase H-bridge LDN configuration can easily be extended to a three-phase configuration, as shown in Figs. 3.5 and 3.6.

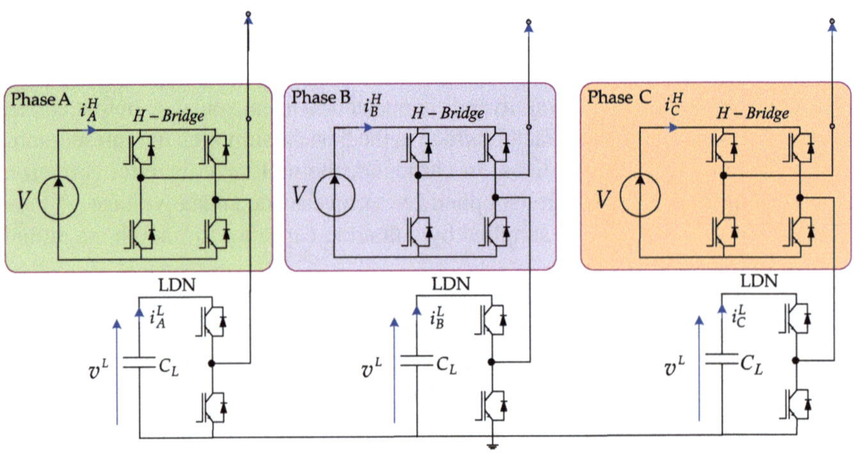

Fig. 3.5 The three-phase H-bridge LDN configuration with three separated dc-links

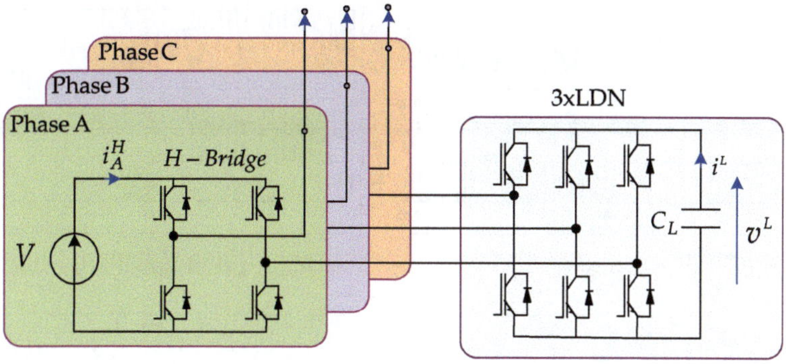

Fig. 3.6 The three-phase H-bridge LDN configuration with a single dc-link

The three-phase scheme depicted in Fig. 3.5 includes three H bridges and three LDN legs, each one is connected to a separate dc-link capacitor. With the advantage of reducing from three to one the dc-link capacitors, and the possibility of mitigating the low-frequency voltage ripple, i.e., reducing the capacitor size, the three LDN legs can be grouped in a single dc-link as shown in Fig. 3.6.

It should be noted that the two above mentioned schemes can be realized with the same number of power switches (i.e. 6 per phase, 18 in all).

3.2.2 *Modulation Principles*

A. **Single-phase**

In order to obtain a proper multilevel output voltages waveform with self-balancing capability (LDN normalized voltage is equal to 1/2), a modulating signal for the LDN is proposed in the following original compact form:

$$u_{ac}^{L} = \begin{cases} |u_{ac}| & \text{for } |u_{ac}| \leq 0.5 \\ 1 - |u_{ac}| & \text{for } |u_{ac}| \geq 0.5 \end{cases} \tag{3.2}$$

For the H-bridge, the modulating signal can be obtained as

$$u_{ac}^{H} = u_{ac} - u_{ac}^{L} \tag{3.3}$$

leading to the following original compact form:

$$u_{ac}^{H} = \begin{cases} u_{ac} - |u_{ac}| & \text{for } |u_{ac}| \leq 0.5 \\ u_{ac} - 1 + |u_{ac}| & \text{for } |u_{ac}| \geq 0.5 \end{cases} \tag{3.4}$$

Modulating signals (3.1)–(3.4), and corresponding normalized instantaneous output voltages of the whole inverter, the H-bridge, and the LDN are depicted in Fig. 3.7 (from top to bottom) in case of modulation indexes 0.5 and 1, respectively.

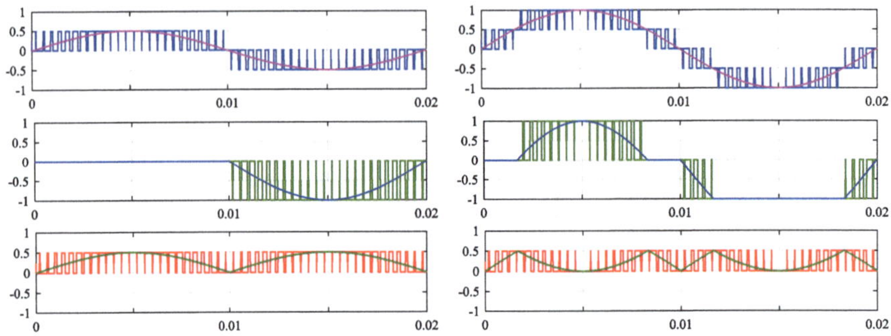

Fig. 3.7 Modulating signals and instantaneous output voltage normalized by V in case of m = 1: total (top), H-Bridge (middle), LDN (bottom) in case of m = 0.5 (left) and m = 1 (right)

Phase disposition modulation is adopted for the two HB carriers. The halved positive carrier is inverted and adopted for modulating the LDN to guarantee the synchronization as depicted in Figs. 3.8 and 3.9 for two cases of modulation indexes. Note that modulating signals coincide with normalized output voltages averaged over the switching period in the linear modulation range ($0 \leq m \leq 1$).

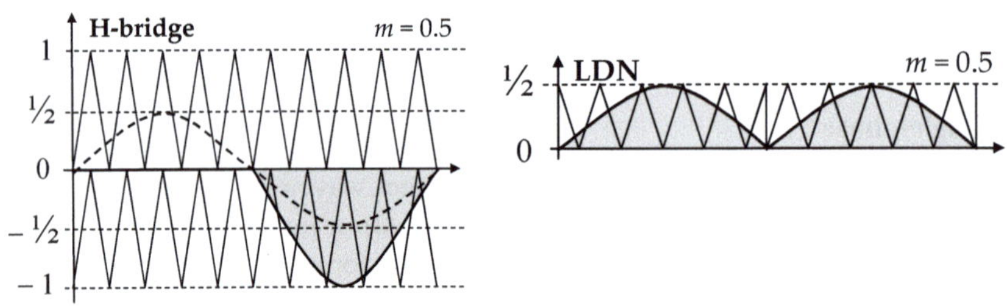

Fig. 3.8 Modulation principle for multilevel PWM: carriers and modulating signals for H-bridge (left) and LDN (right) in case of m = 0.5

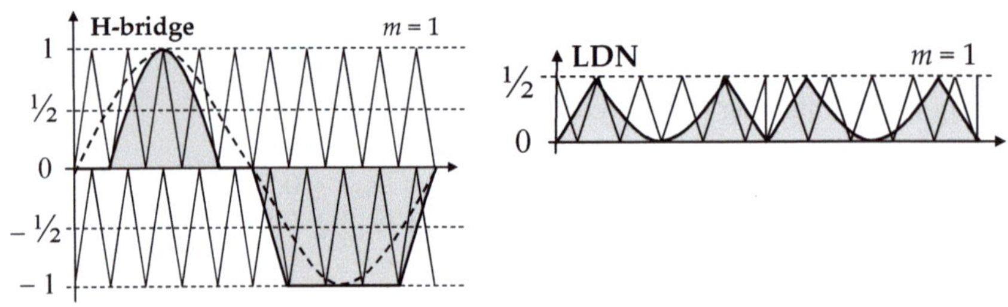

Fig. 3.9 Modulation principle for multilevel PWM: carriers and modulating signals for H-bridge (left) and LDN (right) in case of m = 1

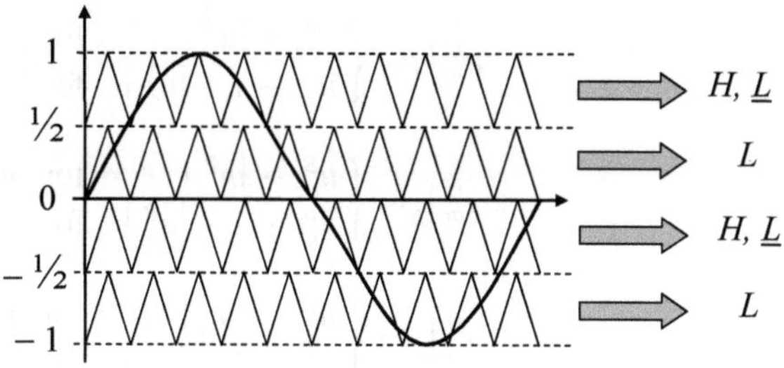

Fig. 3.10 Modulation principle for multilevel PWM

As an alternative way to modulate the proposed LDN multilevel inverter, a level shifted modulation can be implemented as shown in Fig. 3.10.

B. **Three-phase**

The traditional method for generating the switching patterns in three-phase inverters with CB-PWM is the use of sinusoidal wave reference and its comparison with a triangular wave carrier signal known as sinusoidal modulation (SPWM).

In case of SPWM, the modulation principle for the single-phase converter can be directly applied to the three phases, and the sinusoidal references signals for phase A, B and C are written as:

$$\begin{cases} u_{ac}^{A} = m \sin(\vartheta) \\ u_{ac}^{B} = m \sin(\vartheta - \frac{2}{3}\pi) \\ u_{ac}^{C} = m \sin(\vartheta + \frac{2}{3}\pi) \end{cases} \tag{3.5}$$

In order to modulate the three phase H-bridge LDN configuration, level shifted modulation (Fig. 3.10) has been used. By utilizing the same modulating signals of the single-phase LDN configuration and considering (3.5), the normalized modulating signals for the LDN cell (*L*) can simply be derived for each phase as:

$$u_{ac}^{L}\big|_{A} = \begin{cases} \left|u_{ac}^{A}\right| & \text{for } \left|u_{ac}^{A}\right| \le 0.5 \\ 1 - \left|u_{ac}^{A}\right| & \text{for } \left|u_{ac}^{A}\right| \ge 0.5 \end{cases} \tag{3.6}$$

$$u_{ac}^{L}\big|_{B} = \begin{cases} \left|u_{ac}^{B}\right| & \text{for } \left|u_{ac}^{B}\right| \le 0.5 \\ 1 - \left|u_{ac}^{B}\right| & \text{for } \left|u_{ac}^{B}\right| \ge 0.5 \end{cases} \tag{3.7}$$

$$u_{ac}^{L}\big|_{C} = \begin{cases} \left|u_{ac}^{C}\right| & \text{for } \left|u_{ac}^{C}\right| \le 0.5 \\ 1 - \left|u_{ac}^{C}\right| & \text{for } \left|u_{ac}^{C}\right| \ge 0.5 \end{cases} \tag{3.8}$$

As for the LDN cell, modulating signals for the H-bridge cell (H) can be written as:

$$u_{ac}^{H}\big|_{A} = \begin{cases} u_{ac}^{A} - \left|u_{ac}^{A}\right| & \text{for } \left|u_{ac}^{A}\right| \leq 0.5 \\ u_{ac}^{A} - 1 + \left|u_{ac}^{A}\right| & \text{for } \left|u_{ac}^{A}\right| \geq 0.5 \end{cases} \tag{3.9}$$

$$u_{ac}^{H}\big|_{B} = \begin{cases} u_{ac}^{B} - \left|u_{ac}^{B}\right| & \text{for } \left|u_{ac}^{B}\right| \leq 0.5 \\ u_{ac}^{B} - 1 + \left|u_{ac}^{B}\right| & \text{for } \left|u_{ac}^{B}\right| \geq 0.5 \end{cases} \tag{3.10}$$

$$u_{ac}^{H}\big|_{C} = \begin{cases} u_{ac}^{C} - \left|u_{ac}^{C}\right| & \text{for } \left|u_{ac}^{C}\right| \leq 0.5 \\ u_{ac}^{C} - 1 + \left|u_{ac}^{C}\right| & \text{for } \left|u_{ac}^{C}\right| \geq 0.5 \end{cases} \tag{3.11}$$

The instantaneous voltages and the modulating signals are shown in Fig. 3.11 for the two relevant cases of m = 0.5 and m = 1.

As known, a common-mode signal c_m can be injected into the individual inverter voltages, leaving a degree of freedom to the modulation strategy in case of three-phase PWM. In this case, (3.5) becomes:

$$\begin{cases} u_{ac}^{A} = m \sin(\vartheta) + c_m(\vartheta) \\ u_{ac}^{B} = m \sin(\vartheta - \tfrac{2}{3}\pi) + c_m(\vartheta) \\ u_{ac}^{C} = m \sin(\vartheta + \tfrac{2}{3}\pi) + c_m(\vartheta) \end{cases} \tag{3.12}$$

The common-mode signal is usually determined in order to maximize the linear modulation index range and/or to reduce the output current ripple.

Various common-mode methods have been developed so far [22, 23]. The typical common-mode waveform used is a SPWM + third harmonic described as:

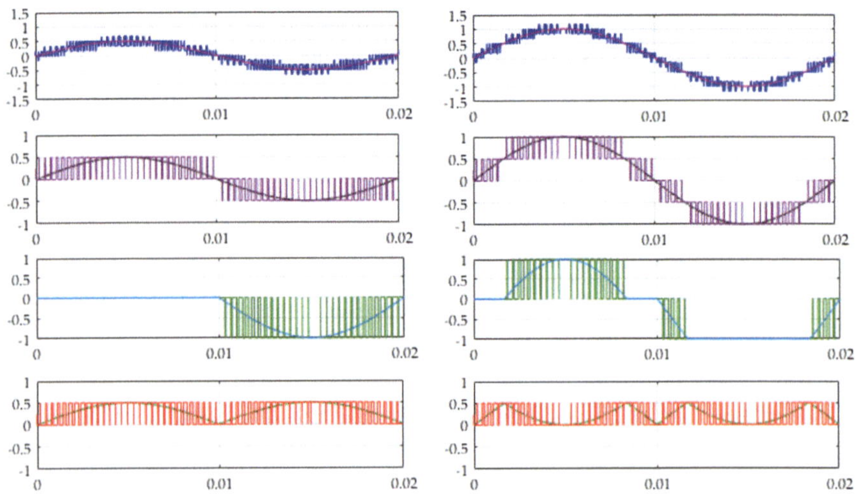

Fig. 3.11 Normalized load voltage, inverter voltage, and individual HB and LDN voltages (top-to bottom) with associated modulating signals in case of m = 0.5 (left) and m = 1 (right), for SPWM

$$c_m = \frac{1}{6} m \sin(3\vartheta) \tag{3.13}$$

In general, the common-mode signal has a triangular waveform (with triple-fundamental frequency), resulting from the centering of the sinusoidal modulating signals (CPWM) known also in the literature as min-max modulation and given by:

$$c_m = -\frac{1}{2}\left[\max\left(u_{ac}^A, u_{ac}^B, u_{ac}^C\right) + \min\left(u_{ac}^A, u_{ac}^B, u_{ac}^C\right)\right] \tag{3.14}$$

The common-mode injection min-max can be approximated as a triangular waveform and can be written as a Fourier series as:

$$c_m \cong \frac{2m}{\pi^2}\left[\sin(3\vartheta) - \frac{1}{9}\sin(9\vartheta)\right] \tag{3.15}$$

Introducing (3.15) in (3.12), leads to:

$$\begin{cases} u_{ac}^A = m\sin(\vartheta) + \frac{2m}{\pi^2}\sin(3\vartheta) - \frac{2m}{9\pi^2}\sin(9\vartheta) \\ u_{ac}^B = m\sin(\vartheta - \frac{2}{3}\pi) + \frac{2m}{\pi^2}\sin(3\vartheta) - \frac{2m}{9\pi^2}\sin(9\vartheta) \\ u_{ac}^C = m\sin(\vartheta + \frac{2}{3}\pi) + \frac{2m}{\pi^2}\sin(3\vartheta) - \frac{2m}{9\pi^2}\sin(9\vartheta) \end{cases} \tag{3.16}$$

By introducing (3.12) in (3.6) and (3.9), the normalized LDN and H-bridge modulating signals of the phase A can be written in case of CPWM as:

$$u_{ac}^L\big|_A = \begin{cases} |m\sin(\vartheta) + c_m(\vartheta)|, & |m\sin(\vartheta) + c_m(\vartheta)| \leq 0.577 \\ 1 - |m\sin(\vartheta) + c_m(\vartheta)|, & |m\sin(\vartheta) + c_m(\vartheta)| \geq 0.577 \end{cases} \tag{3.17}$$

$$u_{ac}^H\big|_A = \begin{cases} m\sin(\vartheta) + c_m(\vartheta) - |m\sin(\vartheta) + c_m(\vartheta)|, & |m\sin(\vartheta) + c_m(\vartheta)| \leq 0.577 \\ m\sin(\vartheta) + c_m(\vartheta) - 1 + |m\sin(\vartheta) + c_m(\vartheta)|, & |m\sin(\vartheta) + c_m(\vartheta)| \geq 0.577 \end{cases} \tag{3.18}$$

The same for phases B and C considering the displacement of 120°.

The normalized load voltage, inverter voltage, and individual HB and LDN voltages with associated modulating signals for the two relevant cases of m = 0.577 and m = 1.15, corresponding to the new modulation boundaries in case of centered PWM are depicted in Fig. 3.12. With the three-phase H-bridge LDN configuration (Fig. 3.6), the load voltage levels can be up to 17 (m = 1.15), resulting from up to 5 levels in the inverter voltage.

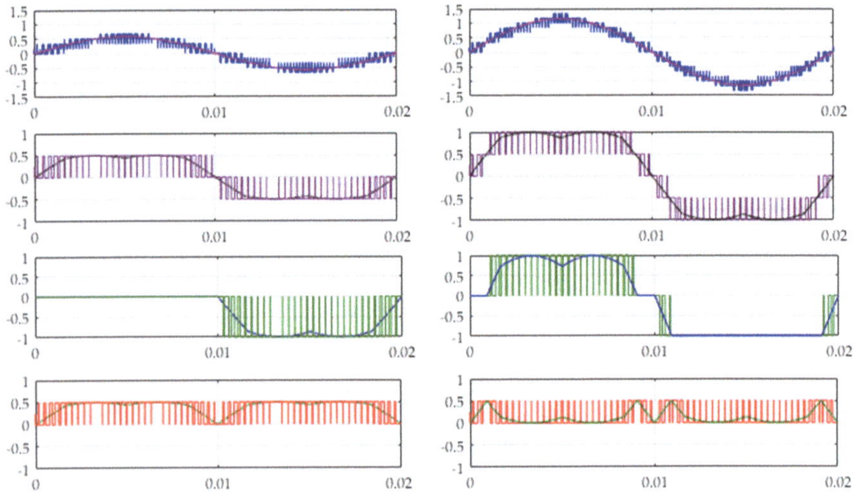

Fig. 3.12 Normalized load voltage, inverter voltage, and individual HB and LDN voltages (top-to bottom) with modulating signals in case of m = 0.577 (left) and m = 1.15 (right), for CPWM

3.3 Harmonic Analysis of Output Voltages

A. **Single-phase**

In order to determine the output voltage harmonics, the symmetry of the LDN voltage can be exploited, according to the waveforms shown in Fig. 3.7. In particular, the harmonic spectrum contains only even harmonics with cosine terms.

The normalized LDN output voltage averaged over the switching period, i.e., the LDN modulating signal (3.2), can be written in terms of harmonics as

$$u_{ac}^{L} = U_0 + \sum_{k=2}^{\infty} U_k \cos(k\,\vartheta) \tag{3.19}$$

where U_0 is the average component and $|U_k|$ is the amplitude of kth harmonic component, being k an even number. The amplitude of these components are calculated analytically as:

$$U_0 = \begin{cases} \frac{2}{\pi} m & \text{for } m \leq 0.5 \\ \frac{2}{\pi}\left[m + \frac{\pi}{2} - \arcsin(1/2m) - \sqrt{4m^2 - 1}\right] & \text{for } m \geq 0.5 \end{cases} \tag{3.20}$$

$$U_k = \begin{cases} -\frac{4}{\pi}\frac{m}{(k^2-1)} & \text{for } m \leq 0.5 \\ \frac{4}{\pi}\left[A_k + B_k\right] & \text{for } m \geq 0.5 \end{cases} \tag{3.21}$$

The expressions of the coefficients A_k and B_k in Eq. (3.21) are given in the Appendix B. Figure 3.13 shows the normalized amplitudes of low-order LDN output voltage harmonics over the whole modulation index range.

According to Eqs. (3.3) and (3.19), the normalized output voltage of the H-bridge is written as:

$$u_{ac}^{H} = -U_0 + m\sin(\vartheta) - \sum_{k=2}^{\infty} U_k \cos(k\vartheta) \tag{3.22}$$

The amplitudes of the harmonics in Eq. (3.22) can be determined considering Eqs. (3.20) and (3.21). Comparing Eq. (3.22) with (3.19) it is evident that, comparing to the LDN, the H-bridge has an additional 1st harmonic component $U_1 = m$. Figure 3.14 shows the normalized amplitudes of low-order LDN output voltage harmonics over the whole modulation index range.

B. **Three-phase**

In case of SPWM and three-phase inverter topology, the output voltage harmonics for each phase are exactly the same as for single-phase, for both H-bridge and LDN cells, as described in (3.19)–(3.22). The amplitudes of low-frequency H-bridge and LDN output voltage harmonics for phase A are exactly the same as those depicted in Figs. 3.13 and 3.14.

In case of CPWM and with reference to phase A, the normalized modulating signals for H-bridge cell (3.6) and LDN cell (3.9) can be expressed in terms of the harmonics as:

$$u_{ac}^{L}\big|_{A} = U_0 + \sum_{k=2}^{\infty} U_k \cos(k\vartheta) \tag{3.23}$$

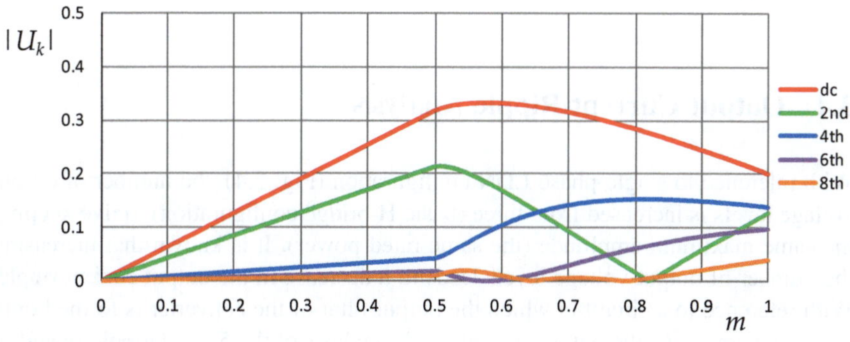

Fig. 3.13 Normalized amplitudes of low-order LDN output voltage harmonics as a function of modulation index

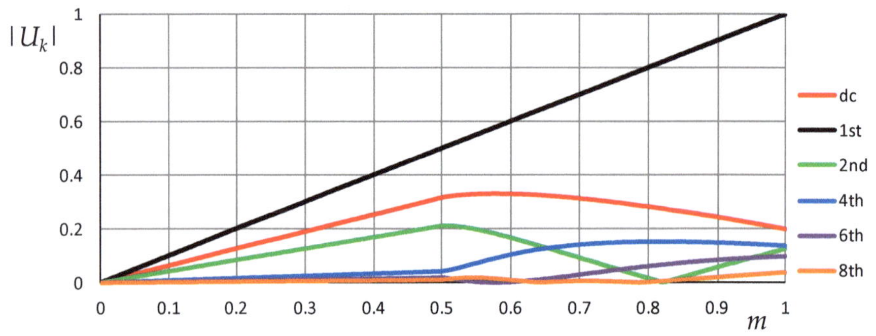

Fig. 3.14 Normalized amplitudes of low-order H-bridge output voltage harmonics as a function of the modulation index

$$u_{ac}^{H}\big|_{A} = -U_0 + m\sin(\vartheta) - \sum_{k=2}^{\infty} U_k \cos(k\vartheta) \tag{3.24}$$

On the basis of (3.12) and (3.15), the amplitudes of U_0 and U_k in case of CPWM have been calculated analytically for $m \leq 0.5$. However, due to the cumbersome calculations for $m > 0.5$, they have been calculated numerically.

For $m \leq 0.5$, Fourier terms of the output voltage harmonics of both H-bridge and LDN cells are calculated analytically as:

$$U_0 = \frac{2}{\pi} m \left[1 + \frac{2}{3\pi^2} - \frac{2}{81\pi^2}\right] \tag{3.25}$$

$$U_k = -\frac{4}{\pi}\frac{m}{(k^2-1)} - \frac{24}{\pi^3}\frac{m}{(k^2-9)} \tag{3.26}$$

Figures 3.15 and 3.16 show the normalized output voltage harmonics of LDN and H-bridge cells, respectively, over the whole modulation index range (calculated numerically for $m > 0.5$, dashed lines), for $k = 0, 2, 4, 6$ and 8 in case of CPWM.

3.4 Output Current Ripple Analysis

With reference to single-phase LDN configuration (Fig. 3.4), the number of output voltage levels is increased from three (basic H-bridge configuration) to five keeping the same maximum amplitude (the same rated power). It is known that increasing the number of output voltage levels leads to a decrease in the output current ripple. With reference to a circuit in which the output filter of the converter is formed only by an inductance L_f, the output current ripple analysis of the 5-level topology and of a basic full-bridge topology driven by unipolar PWM are reported in [13, 24]. In

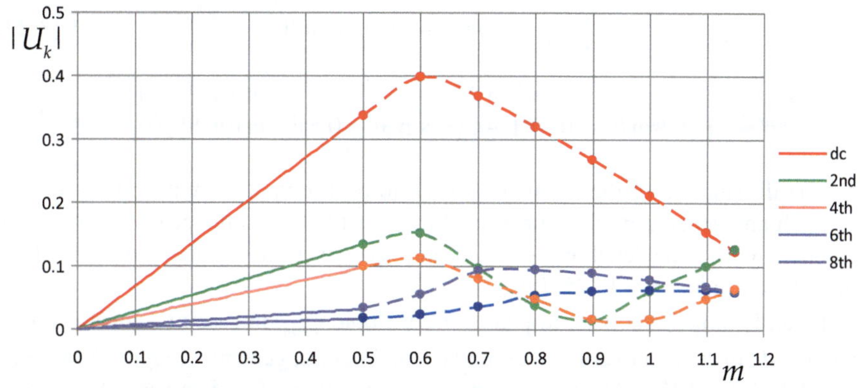

Fig. 3.15 Amplitudes of low-frequency LDN output voltage harmonics as a function of the modulation index in case of CPWM. For $m > 0.5$ (dashed line) it has been calculated numerically

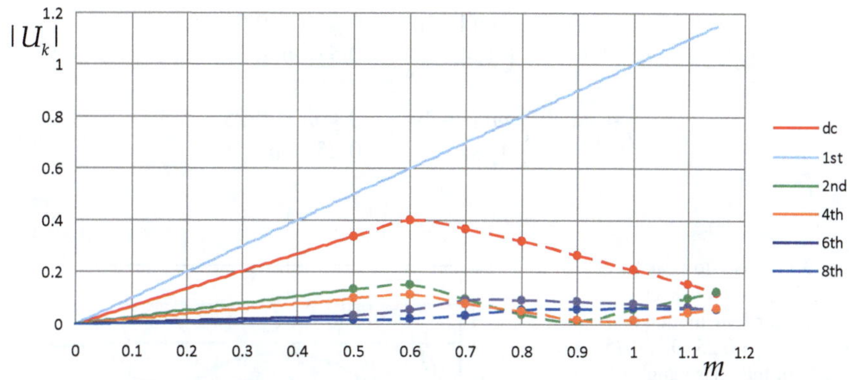

Fig. 3.16 Amplitudes of low-frequency H-bridge output voltage harmonics as a function of the modulation index in case of CPWM. For $m > 0.5$ (dashed line) it has been calculated numerically

particular, the peak-to-peak current ripple $\tilde{i}_{ac,pp}$ for both 3-level (3L) and 5-level (5L) inverter is normalized as follows:

$$\tilde{i}^{3L}_{ac,pp} = \frac{T_{sw}V}{L_f} r^{3L}_{ac,pp} \tag{3.27}$$

$$\tilde{i}^{5L}_{ac,pp} = \frac{T_{sw}V}{L_f} r^{5L}_{ac,pp} \tag{3.28}$$

being T_{sw} the switching period, V the dc supply of the inverter, and $r_{ac,pp}$ the normalized output current ripple of 3-level (3L) and 5-level (5L) inverter, respectively:

$$r_{ac,pp}^{3L} = m(1 - m\sin\vartheta)\sin\vartheta \tag{3.29}$$

$$r_{ac,pp}^{5L} = \begin{cases} m|\sin\vartheta|(1 - 2m|\sin\vartheta|) & m\,|\sin\vartheta| \le 0.5 \\ (m|\sin\vartheta| - 0.5)[1 - 2(m|\sin\vartheta| - 0.5)] & m\,|\sin\vartheta| \ge 0.5 \end{cases} \tag{3.30}$$

The distribution of the normalized peak-to-peak output current ripple over half of the fundamental period of three-level inverter (3.29) and the proposed five-level inverter (3.30) is depicted in Figs. 3.17 and 3.18, respectively, for four cases of the modulation index m = [0.25, 0.5, 0.75, 1].

It can be noted that peak-to-peak current ripple $\tilde{i}_{ac,pp}^{3L}$ in case of 3-level inverter, has a wide excursion over the half fundamental period, generally ranging between 0 and 0.25 (Fig. 3.17). However, it ranges between 0 and 0.125 in case of five-level inverter (Fig. 3.18).

The maximum of the normalized peak-to-peak current ripple for three and five-level inverter is analytically obtained, respectively, as follows:

$$r_{ac}^{3L}\Big|^{\max} = \begin{cases} m(1 - m); & 0 \le m \le 0.5 \\ 1/4; & 0.5 \le m \le 1 \end{cases} \tag{3.31}$$

$$r_{ac}^{5L}\Big|^{\max} = \begin{cases} m(1 - 2m); & 0 \le m \le 0.25 \\ 1/8; & 0.25 \le m \le 1 \end{cases} \tag{3.32}$$

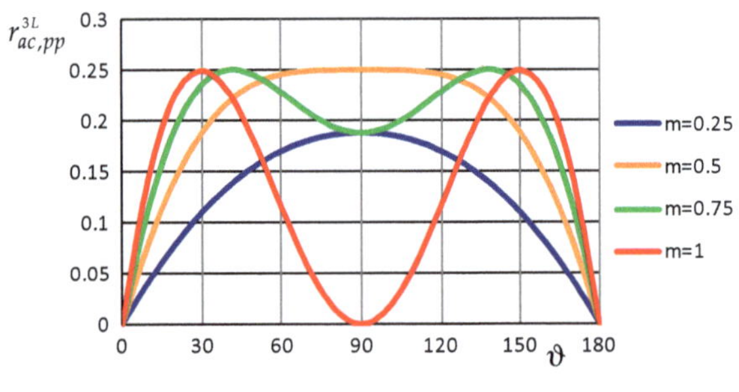

Fig. 3.17 Peak-to-peak current ripple of 3-level inverter for four modulation indexes, m = 0.25, 0.5, 0.75 and 1, in the phase angle range [0, 180°]

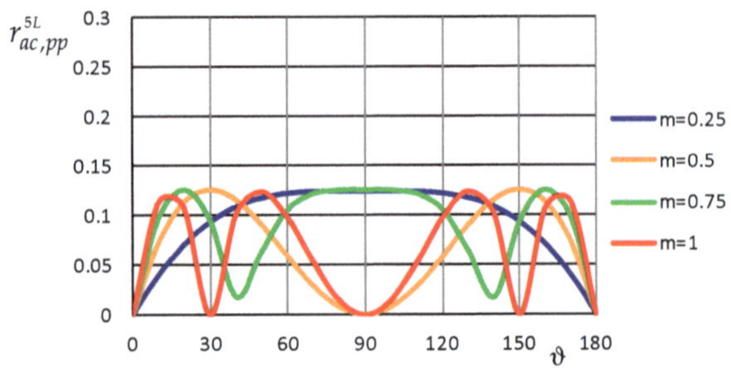

Fig. 3.18 Peak-to-peak current ripple of 5-level inverter for four modulation indexes, m = 0.25, 0.5, 0.75 and 1, in the phase angle range [0, 180°]

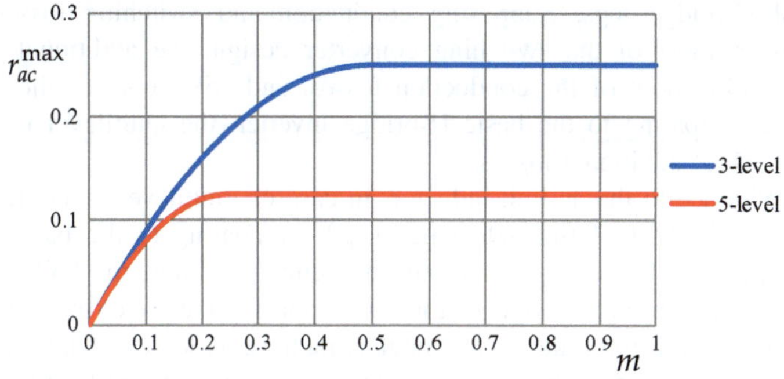

Fig. 3.19 Maximum of normalized peak-to-peak output current ripple as a function of modulation index

In order to show the behavior of the maximum peak-to-peak output current ripple as a function of modulation index, Fig. 3.19 represents the normalized functions r_{ac}^{max} defined by (3.31) and (3.32) for the 3-level and 5-level inverter.

It is clear, as expected, that the maximum current ripple for a five-level inverter is almost the half of the maximum current ripple compared to a three-level inverter, except for the low modulation indexes. The output current ripple decreases with the increase of the output voltage levels.

3.5 Discussion

The considered multilevel conversion scheme (Fig. 3.4) can be compared with the basic H-bridge conversion scheme from the point of view of the overall efficiency. In particular, a comparative estimation of power losses can be carried out introducing some simplifying assumptions. The following considerations can be also applied to each phase of the three-phase multilevel configuration (Fig. 3.6).

Generally speaking, the single-phase multilevel converter itself has 1 additional leg (with 2 additional power switches), i.e. 3 legs instead of 2, comparing to the basic H-bridge configuration. This is leading to an increase of conduction and switching losses, being the main well-known disadvantage of multilevel configurations. On the other hand, the reduced harmonic distortion in multilevel output voltage (inter-level voltage excursion is the half passing from 3 to 5 levels) makes possible to reduce the inductance (L_f) of the ac-link inductor to obtain the same grid current distortion, saving inductor losses, being this one of the known advantages of multilevel configurations.

With reference to the multilevel inverter losses, being the LDN leg in series with the two H-bridge legs, they share the same current, so each leg has the same conduction losses. Being the steady-state voltage the half in the LDN leg, the switching losses in the LDN leg are the half compared to the switching losses in the individual H-bridge legs. Supposing conduction and switching losses equally shared, as in most of the switching converter design, the additional LDN leg introduces 50% more of the conduction losses, and 25% more of the switching losses. So, comparing to the basic H-bridge inverter, the multilevel inverter has almost 37.5% of additional losses.

With reference to the ac-link inductor, in case of multilevel inverter, it can be designed for the half of the inductance ($L_f/2$) comparing to the basic H-bridge scheme (L_f), since the voltage harmonic distortion is almost the half, leading to almost the same current harmonic distortion. Supposing the use of the same amount of copper for the reactor winding, and neglecting the core losses (if any), the copper losses are reduced to 50% being the winding resistance reduced to the half (turns are $\sqrt{2}$ times less and wire cross area can be $\sqrt{2}$ times more to have the same copper weight), and the ac-link reactor current the same. The same output current THD can be obtained by reducing the switching frequency (reducing the switching losses) if the same output filter is adopted.

All in all, a real benefit in terms of efficiency can be experienced in case of the multilevel inverter (H-bridge plus LDN) compared to basic inverter (only H-bridge) especially in case of overall conversion system design with lower inverter losses (that are increased 37.5%) compared to ac-link inductor losses (that are decreasing 50%).

References

1. B. Reznikov, M. Srndovic, Y.L. Familiant, G. Grandi, A. Ruderman, Simple time averaging current quality evaluation of a single-phase multilevel PWM inverter. IEEE Trans. Ind. Electron. **63**(6), 3605–3615 (2016)
2. S. Daher, J. Schmid, F.L.M. Antunes, Multilevel inverter topologies for stand-alone PV systems. IEEE Trans. Ind. Electron. **55**(7), 2703–2712 (2008)
3. D.A. Ruiz-Caballero, R.M. Ramos-Astudillo, S.A. Mussa, M.L. Heldwein, Symmetrical hybrid multilevel DCAC converters with reduced number of insulated DC supplies. IEEE Trans. Ind. Electron. **57**(7), 2307–2314 (2010)
4. F.A. Dragonas, G. Neretti, P. Sanjeevikumar, G. Grandi, High-voltage high-frequency arbitrary waveform multilevel generator for DBD plasma actuators. IEEE Trans. Ind. Appl. **51** (4), 3334–3342 (2015)
5. S.K. Chattopadhyay, C. Chakraborty, A new multilevel inverter topology with self-balancing level doubling network. IEEE Trans. Ind. Electron. **61**(9), 4622–4631 (2014)
6. H. Belkamel, S. Mekhilef, A. Masaoud, M.A. Naeim, Novel three-phase asymmetrical cascaded multilevel voltage source inverter. IET Power Electron. **6**(8), 1696–1706 (2013)
7. S.S. Fazel, S. Bernet, D. Krug, K. Jalili, Design and comparison of 4-kV series-connected H-bridge multilevel converters. Design **43**(4), 1032–1040 (2007)
8. P. Liu, S. Duan, C. Yao, C. Chen, A double modulation wave CBPWM strategy providing neutral-point voltage oscillation elimination and CMV reduction for three-level NPC inverters. IEEE Trans. Ind. Electron. **65**(1), 16–26 (2017)
9. C.L. Xia, G.Z. Zhang, Y. Yan, X. Gu, T.N. Shi, X.N. He, Discontinuous space vector PWM strategy of neutral-point-clamped three-level inverters for output current ripple reduction. IEEE Trans. Power Electron. **32**(7), 5109–5121 (2017)

10. A. Abdelhakim, P. Mattavelli, G. Spiazzi, Three-phase three-level flying capacitors split-source inverters: analysis and modulation. IEEE Trans. Ind. Electron. **64**(6), 4571–4580 (2017)
11. Y. Sato, M. Iimura, Y. Dodo, H. Obara, A study on minimum required capacitance in flying capacitor multilevel converters for grid-connected applications, in *2015 IEEE Energy Conversion Congress and Exposition ECCE 2015*, no. 22560264, pp. 3502–3507, 2015
12. B.M. Hoosh, R. Zaimeddine, T.M. Undeland, Comparison of harmonics and common mode voltage in NPC and FLC multilevel converters, in *Proceeding of EPE-PEMC 2010—14th International Electronics and Motion Control Conference* (2010), pp. 158–161
13. G. Buticchi, E. Lorenzani, G. Franceschini, A five-level single-phase grid-connected converter for renewable distributed systems. IEEE Trans. Ind. Electron. **60**(3), 906–918 (2013)
14. S.K. Chattopadhyay, C. Chakraborty, A new asymmetric multilevel inverter topology suitable for solar PV applications with varying irradiance. IEEE Trans. Sustain. Energy **3029**(99), 1–1 (2017)
15. S. Mariethoz, Design and control of high-performance modular hybrid asymmetrical cascade multilevel inverters. IEEE Trans. Ind. Appl. **50**(6), 4018–4027 (2014)
16. Y. Zhang, L. Sun, An efficient control strategy for a five-level inverter comprising flying-capacitor asymmetric H-bridge. IEEE Trans. Ind. Electron. **58**(9), 4000–4009 (2011)
17. N.A. Rahim, J. Selvaraj, Multistring five-level inverter with novel PWM control scheme for PV application. IEEE Trans. Ind. Electron. **57**(6), 2111–2123 (2010)
18. D. Floricau, F. Richardeau, New multilevel converters based on stacked commutation cells with shared power devices. IEEE Trans. Ind. Electron. **58**(10), 4675–4682 (2011)
19. K. Gnanasambandam, A.K. Rathore, A. Edpuganti, D. Srinivasan, J. Rodriguez, Current-fed multilevel converters: an overview of circuit topologies, modulation techniques, and applications. IEEE Trans. Power Electron. **32**(5), 3382–3401 (2017)
20. A. Ovalle, M.E. Hernández, G.A. Ramos, A flexible nonorthogonal-reference-frame-based SVPWM framework for multilevel inverters. IEEE Trans. Power Electron. **32**(6), 4925–4938 (2017)
21. A.M. Hava, R.J. Kerkman, T.A. Lipo, Carrier-based PWM-VSI overmodulation strategies: analysis, comparison, and design. IEEE Trans. Power Electron. **13**(4), 674–689 (1998)
22. W. Li, X. He, J. Hu, X. Yuan, W. Zhou, Y. Dong, H. Yang, Common-mode voltage injection-based nearest level modulation with loss reduction for modular multilevel converters. IET Renew. Power Gener. **10**(6), 798–806 (2016)
23. J. Huang, Q. Liu, X. Wang, K. Li, A carrier-based modulation scheme to reduce the third harmonic component of common-mode voltage in a three-phase inverter under high DC voltage utilization. IEEE Trans. Ind. Electron. **46**, 1–1 (2017)
24. G. Grandi, J. Loncarski, O. Dordevic, Analysis and comparison of peak-to-peak current ripple in two-level and multilevel PWM inverters. IEEE Trans. Ind. Electron. **62**(5), 2721–2730 (2015)

Authored Papers

25. M. Hammami, G. Grandi, Single-phase multilevel PV generation system with improved ripple correlation control MPPT algorithm. Energies **10**(12), 2037 (2017)

Chapter 4
Analysis of DC-Link Current and Voltage Ripple: Single-Phase Configuration

4.1 Introduction

A lot of analyses related to PWM techniques have been published, but they were concentrated mostly on the inverter output voltage and current characteristics, for both single-phase and three-phase configurations. A complete analysis of switching losses, dc link voltage harmonics, and inverter input/output current harmonics in three-phase PWM converter system is given in [1]. The analysis of the output current ripple amplitude, introducing simple and effective expressions for determining the maximum amplitude of the peak-to-peak current ripple over the fundamental period, for two-level and multilevel inverters is presented in [2, 3].

However, only a few papers have investigated the inverter characteristics at the input (dc-link) side. This analysis focuses on the dc-link capacitor design, which is thought to be one of the critical components in voltage source inverters due to its cost and susceptibility to temperature degradation [4]. Sufficient attention should be paid to the selection and dimensioning of the dc-bus capacitor in order to prevent overheating and extend the lifetime.

The analysis of the dc-link current and voltage ripple in single-phase H-bridge inverters have been presented in [5, 6]. In case of asymmetric single-phase multilevel inverters, dc-link current and voltage for the FCaHB and NPCaHB configurations have been investigated in [7, 8].

With reference to Fig. 4.1, a detailed analysis of the dc-link current and voltage ripple for the H-bridge and LDN cells is developed.

In particular, the peak-to-peak dc-link current and voltage ripple amplitudes are analytically calculated over the fundamental period as a function of the modulation index for both the low-frequency and switching ripple components. Reference is made to sinusoidal output current with unity power factor, representing the working condition of most of the grid-connected inverters. Maximum peak-to-peak values of

M. Hammami, *Level Doubling Network and Ripple Correlation Control MPPT Algorithm for Grid-Connected Photovoltaic Systems*, Springer Theses,
https://doi.org/10.1007/978-3-030-10492-4_4

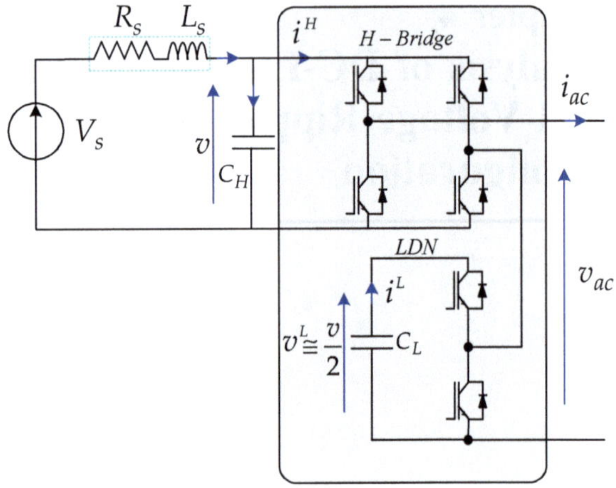

Fig. 4.1 Block diagram of a single-phase CaHB multilevel inverter (H-bridge and LDN)

both low frequency and switching frequency voltage ripple components are effectively adopted to design the capacitors of both H-bridge and LDN cells basing on the desired dc-link voltage ripple requirements.

4.2 Analysis of Input Currents

The superscripts L and H are introduced for LDN and H-bridge cells, respectively. For both cells, the input current can be written as:

$$i^L = \bar{i}^L + \Delta i^L = I^L + \tilde{i}^L + \Delta i^L \tag{4.1}$$

$$i^H = \bar{i}^H + \Delta i^H = I^H + \tilde{i}^H + \Delta i^H \tag{4.2}$$

being $\bar{i}^L$ and $\bar{i}^H$ the averaged input current over the switching period of LDN and H-bridge cells, respectively, consisting of dc (average) components I^L and I^H, and low frequency components, $\tilde{i}^L$ and $\tilde{i}^H$. Δi^L and Δi^H represent the switching current ripple components.

4.2.1 Low-Frequency (Averaged) Current Components

The multilevel inverter (Fig. 4.1) is supplied with a constant voltage source V_s knowing that the steady-state LDN average dc-link voltage is automatically kept around one half of the dc-link voltage of the H-bridge, i.e. $v^L \approx v/2$, as proved in [5] for symmetric LDN operations.

Both H-bridge and LDN dc-link voltages can be considered almost constant. Being V and $V/2$ the corresponding average components (over the fundamental period), the normalized H-bridge and LDN voltages become 1 and 1/2, respectively. With this assumption, and considering the averaged quantities over the switching period, the power balance for LDN and H-bridge cells can be written as:

$$\begin{cases} \frac{1}{2}\bar{i}^L = \bar{u}_{ac}^L \, \bar{i}_{ac} \\ \bar{i}^H = \bar{u}_{ac}^H \, \bar{i}_{ac} \end{cases} \tag{4.3}$$

being u_{ac}^L and u_{ac}^H the normalized modulating signals for LDN and H-bridge cells expressed as:

$$u_{ac}^L = \begin{cases} m\,|\sin\vartheta| & \text{for } m|\sin\vartheta| \leq 0.5 \\ 1 - m\,|\sin\vartheta| & \text{for } m|\sin\vartheta| \geq 0.5 \end{cases} \tag{4.4}$$

$$u_{ac}^H = \begin{cases} m\,\sin\vartheta - m\,|\sin\vartheta| & \text{for } m\,|\sin\vartheta| \leq 0.5 \\ m\,\sin\vartheta - 1 + m\,|\sin\vartheta| & \text{for } m\,|\sin\vartheta| \geq 0.5 \end{cases} \tag{4.5}$$

For all the following analytical developments, reference is made to a sinusoidal output current with unity power factor, which is a reasonable assumption for most of the grid-connected applications. Hence, the instantaneous output current is:

$$i_{ac} \approx \bar{i}_{ac} = I_{ac}\,\sin\vartheta \tag{4.6}$$

where I_{ac} is the output current amplitude.

Replacing (4.4), (4.5) and (4.6) in (4.3), the averaged input currents are calculated analytically as:

$$\bar{i}^L = \bar{i}^L = \begin{cases} 2\,m\,I_{ac}\,\sin\vartheta\,|\sin\vartheta|, & m\,|\sin\vartheta| \leq 0.5 \\ 2\,I_{ac}\,\sin\vartheta\,(1 - m\,|\sin\vartheta|), & m\,|\sin\vartheta| \geq 0.5 \end{cases} \tag{4.7}$$

$$\bar{i}^H = \begin{cases} I_{ac}\,\sin\vartheta\,(m\,\sin\vartheta - m\,|\sin\vartheta|), & m\,|\sin\vartheta| \leq 0.5 \\ I_{ac}\,\sin\vartheta\,(m\,\sin\vartheta - 1 + m\,|\sin\vartheta|), & m\,|\sin\vartheta| \geq 0.5 \end{cases} \tag{4.8}$$

It should be noted that (4.7) and (4.8) represent the low-frequency components for LDN and H-bridge cells, respectively. Similar results have been obtained analytically in [7] for the neutral point capacitors current in case of a NPCaHB converter, and numerically in [8] in the FCaHB case.

Furthermore, the harmonic spectra of LDN and H-bridge input currents are analytically calculated as (see Appendix B):

$$\bar{i}^L = \left[(2U_0 - U_2)\sin\vartheta - \sum_{n=3}^{\infty}(U_{n+1} - U_{n-1})\sin n\vartheta\right] I_{ac} \tag{4.9}$$

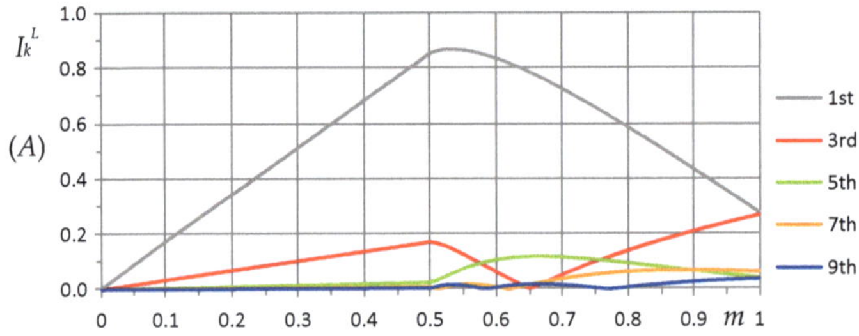

Fig. 4.2 Amplitudes of low-order LDN input current harmonics for sinusoidal unity output current as a function of the modulation index

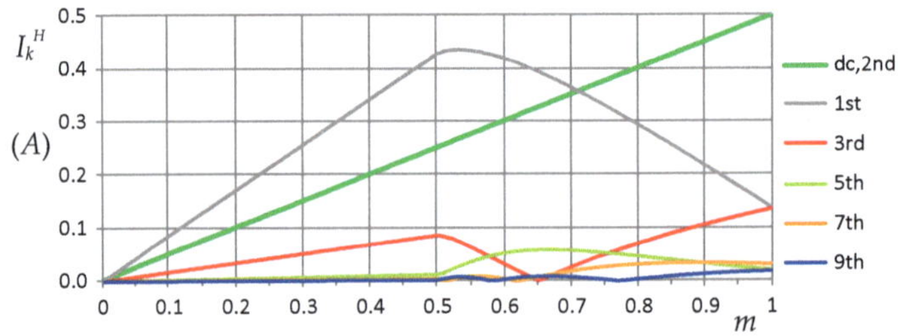

Fig. 4.3 Amplitudes of low-order H-bridge input current harmonics for sinusoidal unity output current as a function of the modulation index

$$\bar{i}^H = \frac{1}{2}\left[m + (2U_0 - U_2)\sin\vartheta - m\cos 2\vartheta - \sum_{n=3}^{\infty}(U_{n+1} - U_{n-1})\sin n\vartheta\right] I_{ac} \tag{4.10}$$

where $n \geq 3$ is odd, and the terms U_0, U_2 and U_n are defined in the Appendix B. Figures 4.2 and 4.3 show the harmonic amplitudes appearing in (4.9) and (4.10) for LDN and H-bridge cells, respectively, as a function of the modulation index.

It is clear that the LDN input current contains just odd harmonics being null the dc component ($I^L = 0$). It should be noted in Fig. 4.3 that the dc and 2nd harmonic components have the same amplitude, proportional to the modulation index. The 1st harmonic has a relevant amplitude, comparable to the amplitude of the 2nd harmonic. Higher order harmonics are quite negligible in the whole modulation range apart the 3rd that can be noticeable around $m = 0.5$ and $m = 1$.

4.2.2 Switching Frequency Current Components

Once the low-frequency components are determined from (4.7) and (4.8) for LDN and H-bridge cells, respectively, the switching current components Δi^L and Δi^H can

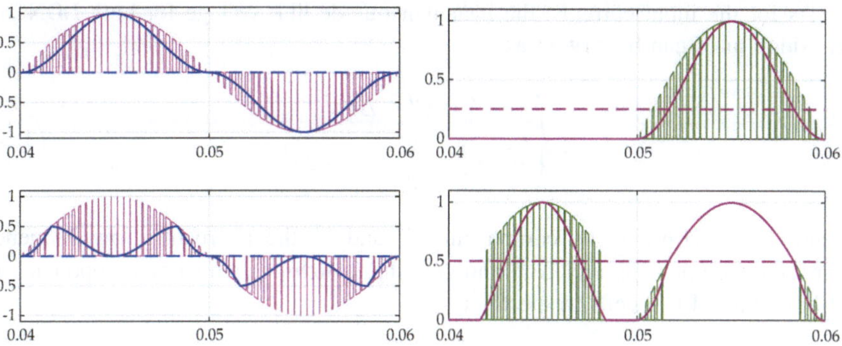

Fig. 4.4 Input current and its averaged counterpart for m = 0.5 (top) and m = 1 (bottom) of the LDN cell (left) and H-bridge cell (right) in case of unity output current (I_{ac} = 1A)

be calculated basing on (4.1) for the LDN cell and basing on (4.2) for H-bridge cell. In particular, focusing on the "on-time" of the upper power switch, the switching ripple components of the input currents can be expressed as:

$$\begin{cases} \Delta i^L = i^L - \bar{i}^L = I_{ac} \sin \vartheta - \bar{i}^L \\ \Delta i^H = i^H - \bar{i}^H = I_{ac} \left| \sin \vartheta \right| - \bar{i}^H \end{cases} \tag{4.11}$$

Introducing (4.7) and (4.8) in (4.11), the aforementioned contributions can be calculated analytically as:

$$\Delta i^L = \begin{cases} I_{ac} \sin \vartheta (1 - 2m \left| \sin \vartheta \right|), & m \left| \sin \vartheta \right| \leq 0.5 \\ I_{ac} \sin \vartheta (-1 + 2m \left| \sin \vartheta \right|), & m \left| \sin \vartheta \right| \geq 0.5 \end{cases} \tag{4.12}$$

$$\Delta i^H = \begin{cases} I_{ac} \{ \left| \sin \vartheta \right| - m \sin \vartheta (\sin \vartheta - \left| \sin \vartheta \right|) \}, & m \left| \sin \vartheta \right| \leq 0.5 \\ I_{ac} \{ \left| \sin \vartheta \right| - \sin \vartheta (m \sin \vartheta - 1 + m \left| \sin \vartheta \right|) \}, & m \left| \sin \vartheta \right| \geq 0.5 \end{cases} \tag{4.13}$$

The instantaneous input current, its low-frequency component corresponding to the averaged value over the switching period, and its dc component are shown in Fig. 4.4 for both LDN (left) and H-bridge cell (right).

The low-frequency components shown in Fig. 4.4 are calculated by (4.7) and (4.8) for LDN and H bridge cells, respectively.

4.3 Analysis of Input Voltages

Basing on the input currents, the analysis can be developed also for the input voltages of both LDN and H-bridge cells. In particular, to ensure a proper design of the dc-link capacitors, the peak-to-peak values of the voltage ripple components should be analytically determined.

As for the input currents, the instantaneous dc-link voltage for both LDN and H-bridge units can be written as:

$$\begin{cases} v = V + \tilde{v}^H + \Delta v^H \\ v^L \cong \dfrac{v}{2} = \dfrac{V}{2} + \tilde{v}^L + \Delta v^L \end{cases} \tag{4.14}$$

being V the H-bridge dc component, $\tilde{v}^H$ and $\tilde{v}^L$ the alternating low-frequency harmonic components and Δv^H and Δv^L the switching frequency components of H-bridge and LDN cells, respectively.

4.3.1 Low-Frequency (Averaged) Voltage Components

A. **LDN cell**

With reference to the LDN cell, the dc component has been proved in [9, 10] to be $V/2$. The low-frequency dc-link voltage component can be calculated analytically by integrating the corresponding input current (4.7) as:

$$\tilde{v}^L = \frac{1}{\omega C_L}\int \tilde{i}^L(\vartheta)\, d\vartheta + K \tag{4.15}$$

where K is an integration constant. Introducing (4.7) in (4.15) and solving the integral, the input voltage becomes:

$$\tilde{v}^L = \frac{I_{ac}}{2f\, C_L}\,\tilde{u}^L \tag{4.16}$$

being $\tilde{u}^L$ the normalized averaged (low frequency) of the LDN input voltage given by

$$\tilde{u}^L = \begin{cases} \frac{m}{\pi}\left(\vartheta - \frac{1}{2}\sin 2\vartheta\right) + K_1, & m\left|\sin\vartheta\right| \le 0.5 \\ -\frac{2}{\pi}\cos\vartheta - \frac{m}{\pi}\left(\vartheta - \frac{1}{2}\sin 2\vartheta\right) + K_2, & m\left|\sin\vartheta\right| \ge 0.5 \end{cases} \tag{4.17}$$

The values of K_1 and K_2 can be determined by setting the continuity of the voltage between the different regions. The waveform of $\tilde{u}^L$ is depicted in Fig. 4.5 in the whole fundamental period, for different modulation indexes.

It can be noticed that the maximum value is always in the middle of the period. As a consequence, the peak-to-peak of the LDN low-frequency voltage ripple component can readily be calculated by setting $\vartheta = \pi$ in (4.17), to yield:

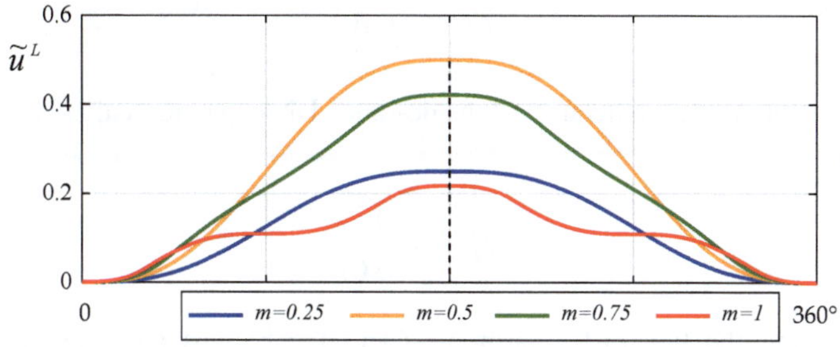

Fig. 4.5 Normalized low-frequency LDN input voltage in fundamental period [0, 360°], for different modulation indices, m = 0.25, 0.5, 0.75 and 1

$$\tilde{u}_{pp}^{L} = \begin{cases} m, & m \leq 0.5 \\ \frac{4}{\pi} m \arcsin(\frac{1}{2m}) - m + \frac{1}{m\pi} \sqrt{4m^2 - 1}, & m \geq 0.5 \end{cases} \quad (4.18)$$

Figure 4.6 shows $\tilde{u}_{pp}^{L}$ as a function of modulation index. The maximum amplitude of normalized peak-to-peak low-frequency voltage ripple is observed around $m \approx 0.55$, and can be analytically calculated from (4.18) as:

$$\tilde{u}_{pp}^{L}\Big|^{\max} = 0.5143 \cong \frac{1}{2} \quad (4.19)$$

Finally, combining (4.16) and (4.19), the maximum amplitude of the low-frequency peak-to-peak voltage ripple can be written as a simple function of output current magnitude and LDN capacitance as:

$$\tilde{v}_{pp}^{L}\Big|^{\max} = \frac{I_{ac}}{4f\,C_L} \quad (4.20)$$

As an alternative, the low-frequency dc-link voltage component can be calculated on the basis of the amplitude of the low-frequency input current component (4.9) and the corresponding dc-link reactance, as follows:

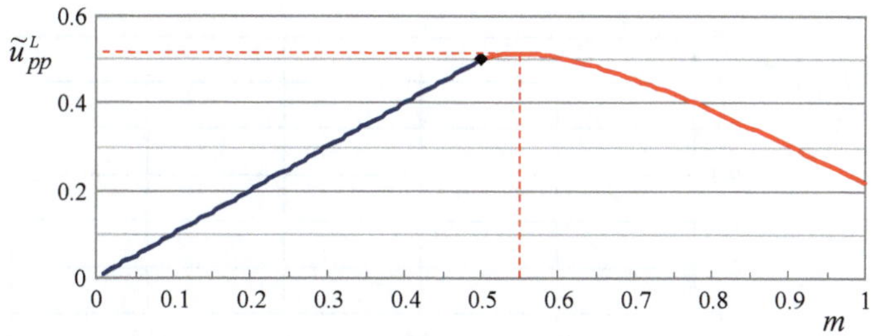

Fig. 4.6 Peak-to-peak value of the normalized low-frequency LDN input voltage versus modulation index

$$V_k^L = Z_k^L \, I_k^L \tag{4.21}$$

being Z_k^L the equivalent reactance of the dc-link LDN capacitor, calculated at the kth harmonic:

$$Z_k^L = \frac{1}{k \, \omega \, C_L} \tag{4.22}$$

Considering the first order harmonic (with the highest amplitude), and combining (4.9), (4.21) and (4.22), the peak-to-peak LDN low-frequency voltage component at the fundamental frequency can be expressed as:

$$V_{1,pp}^L = \frac{2\,[2U_0 - U_2]}{2\,\pi f \; C_L} = \frac{I_{ac}}{2f\,C_L} \; U_{1,pp}^L \tag{4.23}$$

being $U_{1,pp}^L$ the normalized LDN input voltage expressed as:

$$U_{1,pp}^L = \frac{2}{\pi}\,[2U_0 - U_2] \tag{4.24}$$

Figure 4.7 presents $U_{1,pp}^L$ as a function of the modulation index. The maximum amplitude of the normalized peak-to-peak voltage ripple at fundamental frequency is:

$$U_{1,pp}^L \Big|^{\max} = 0.55 \;\cong\; \frac{1}{2} \tag{4.25}$$

Finally, the maximum amplitude of the peak-to-peak voltage ripple at the fundamental frequency leads to the same results as (4.20):

$$V_{1,pp}^L \Big|^{\max} \cong \frac{I_{ac}}{4f\,C_L} \tag{4.26}$$

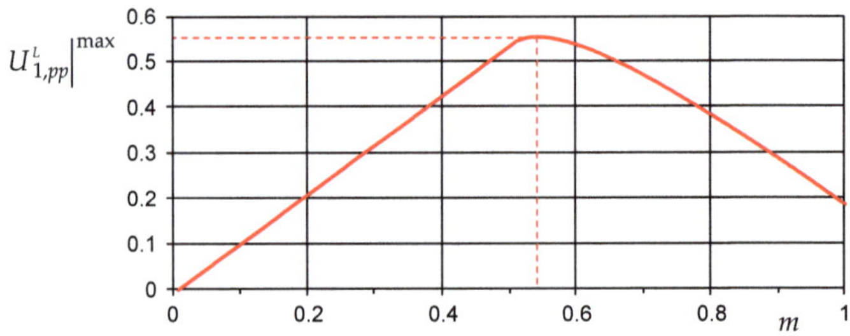

Fig. 4.7 Peak-to-peak value of the normalized LDN input voltage at the fundamental frequency versus the modulation index m

B. **H-bridge cell**

The dc component of the H-bridge cell is equal to V. The amplitudes of the low-frequency input voltage components at kth harmonic V_k^H can be calculated on the basis of the amplitudes of the low-frequency input current components (4.10), and the dc-link equivalent impedance as:

$$V_k^H = Z_k^H \, I_k^H \tag{4.27}$$

being Z_k^H the parallel between the reactance of the dc-link capacitor C_H, and the equivalent input resistance R_s, calculated at the kth harmonic order:

$$Z_k^H = \frac{R_s}{\sqrt{k^2 \, R_s^2 \, \omega^2 \, C_H^2 + 1}} \tag{4.28}$$

Considering realistic parameters of a photovoltaic module, the equivalent series resistance R_s is always larger than $(1/k \; C_H)$. In this case, Eq. (4.28) becomes:

$$Z_k^H = \frac{1}{k \, \omega \, C_H} \tag{4.29}$$

Introducing (4.29) in (4.27) and considering (4.10), the amplitudes of the low-frequency input voltage components at the kth harmonic become:

$$V_k^H = \frac{1}{k \, \omega \, C_H} I_k^H \tag{4.30}$$

Considering only the first and second harmonic order, the voltage amplitudes can be expressed as:

$$V_1^H = \frac{1}{\omega \, C_H} \frac{[2U_0 - U_2]}{2} I_{ac} = \frac{1}{2\pi f \, C_H} \frac{[2U_0 - U_2]}{2} I_{ac} \tag{4.31}$$

$$V_2^H = \frac{1}{2 \, \omega \, C_H} \frac{m}{2} I_{ac} = \frac{1}{4\pi f \, C_H} \frac{m}{2} I_{ac} \tag{4.32}$$

4.3.2 *Switching Frequency Voltage Components*

A. **LDN cell**

Considering the sinusoidal PWM principle, the "on-time" interval Δt^L for the LDN cell is calculated on the basis of the normalized LDN modulating signal described in Chap. 3 by Eq. (3.2) as:

$$\Delta t^L = \begin{cases} 2\,m\,|\sin\vartheta|\,T_{sw}; & m\,|\sin\vartheta| \le 0.5 \\ 2\,(1 - m\,|\sin\vartheta|)\,T_{sw}; & m\,|\sin\vartheta| \ge 0.5 \end{cases} \tag{4.33}$$

The corresponding dc voltage variations (peak-to-peak) over the sub-periods $[0 - \Delta t^L]$, can be expressed as

$$\Delta v^L_{pp} = \frac{1}{C_L}\int_0^{\Delta t^L} \Delta i^L\, dt = \frac{1}{C_L}\Delta i^L\,\Delta t^L \tag{4.34}$$

By introducing in (4.34) the on-time interval Δt^L given by (4.33) and the corresponding current Δi^L given by (4.12) and supposed constant within the interval, the peak-to-peak switching ripple amplitude of the input LDN cell voltage becomes:

$$\Delta v^L_{pp} = \frac{I_{ac}}{f_{sw}\,C_L}\,\Delta u^L_{pp} \tag{4.35}$$

being Δu^L_{pp} the normalized peak-to-peak voltage switching ripple given by

$$\Delta u^L_{pp} = \begin{cases} 2\,m\,\sin^2\vartheta\,(1 - 2\,m\,|\sin\vartheta|), & m\,|\sin\vartheta| \le 0.5 \\ 2\,|\sin\vartheta|\,(2\,m\,|\sin\vartheta| - 1)\,(1 - m\,|\sin\vartheta|), & m\,|\sin\vartheta| \ge 0.5 \end{cases} \tag{4.36}$$

Figure 4.8 presents the distribution of the normalized peak-to-peak ripple amplitude given by (4.36), over the fundamental period ϑ = [0, 360°], for four cases of the modulation index m =[0.25, 0.5, 0.75, 1]. It can be noted that it is always symmetrically distributed.

B. **H-bridge cell**

As for the LDN unit, the "on-time" interval Δt^H for the H-bridge cell is calculated on the basis of the normalized H-bridge modulating signal described in Chap. 3 by Eq. (3.4) as:

$$\Delta t^H = \begin{cases} (m\,\sin\vartheta - |m\,\sin\vartheta|)T_{sw}; & m\,|\sin\vartheta| \le 0.5 \\ [m\,\sin\vartheta - (1 - |m\,\sin\vartheta|)]\,T_{sw}; & m\,|\sin\vartheta| \ge 0.5 \end{cases} \tag{4.37}$$

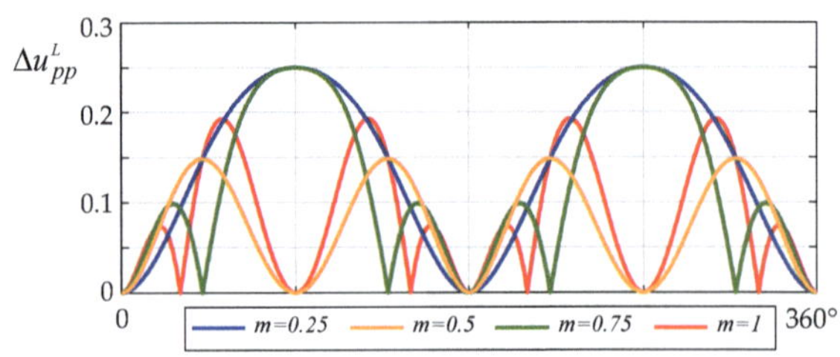

Fig. 4.8 Normalized peak-to-peak switching ripple of the LDN input voltage in the fundamental period [0, 360°], for different modulation indices, m = 0.25, 0.5, 0.75 and 1

The corresponding dc voltage variations (peak-to-peak) over the sub-periods $[0 - \Delta t^H]$, can be expressed as

$$\Delta v_{pp}^H = \frac{1}{C_H} \int_0^{\Delta t^H} \Delta i^H \, dt = \frac{1}{C_H} \Delta i^H \, \Delta t^H \tag{4.38}$$

The peak-to-peak switching ripple amplitude of the input H-bridge voltage can be calculated by introducing (4.13) and (4.37) in (4.38):

$$\Delta v_{pp}^H = \frac{I_{ac}}{f_{sw} C_H} \, \Delta u_{pp}^H \tag{4.39}$$

being Δu_{pp}^H the normalized peak-to-peak voltage switching given by:

$$\Delta u_{pp}^H = \begin{cases} [|\sin\vartheta| - \sin\vartheta\,(m\,\sin\vartheta - m\,|\sin\vartheta|)]\,[m\,\sin\vartheta - |m\,\sin\vartheta|], & m\,|\sin\vartheta| \le 0.5 \\ [|\sin\vartheta| - \sin\vartheta\,(m\,\sin\vartheta - 1 + m\,|\sin\vartheta|)]\,[m\,\sin\vartheta - (1 - |m\,\sin\vartheta|)], & m\,|\sin\vartheta| \ge 0.5 \end{cases} \tag{4.40}$$

Figure 4.9 presents the distribution of the normalized peak-to-peak ripple amplitude given by (4.40), over the fundamental period ϑ = [0, 360°], for four cases of the modulation index m = [0.25, 0.5, 0.75, 1].

In order to estimate the maximum voltage switching ripple amplitude in the fundamental period for both LDN and H-bridge cells, the following expressions in four ranges of m are analytically calculated from (4.36) and (4.40):

$$\left.\Delta u_{pp}^H\right|^{\max} = \left.\Delta u_{pp}^L\right|^{\max} = \begin{cases} 2m\,(1-2m); & 0 \le m \le \frac{1}{3} \\ \frac{2}{27m}; & \frac{1}{3} \le m \le 0.575 \\ 2\,(2m-1)\,(1-m); & 0.575 \le m \le \frac{3+\sqrt{3}}{6} \\ \frac{\sqrt{3}}{9m}; & \frac{3+\sqrt{3}}{6} \le m \le 1 \end{cases} \tag{4.41}$$

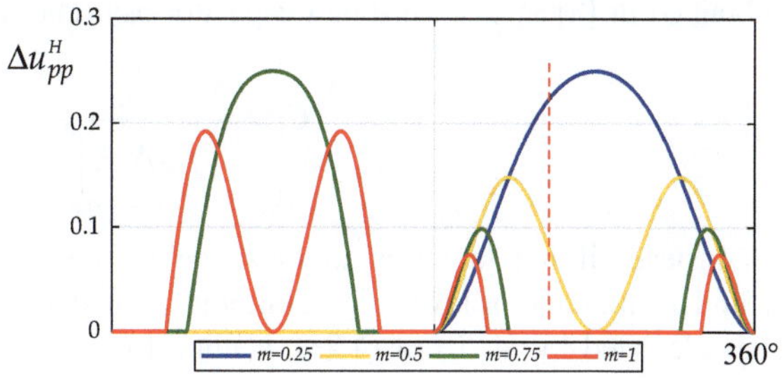

Fig. 4.9 Normalized peak-to-peak voltage switching ripple of H-bridge in fundamental period [0, 360°], for different modulation indices, m = 0.25, 0.5, 0.75 and 1

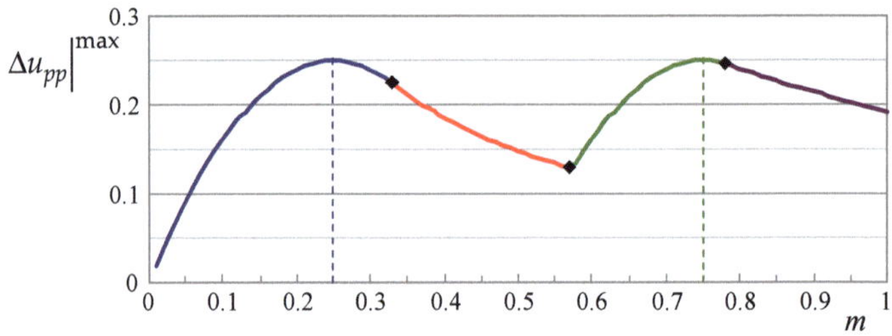

Fig. 4.10 Max of normalized peak-to-peak switching ripple of both H-bridge and LDN input voltage cells versus modulation index

The maximum of the normalized peak-to-peak voltage ripple as a function of m is depicted in Fig. 4.10. It can be noted that it ranges between 0 (m = 0) and 1/4 for both m = 1/4 and m = 3/4. The absolute maximum for both LDN and H-bridge cells become:

$$\Delta v_{pp}^{L}\Big|^{\max} = \frac{I_{ac}}{4 f_{sw} C_L} \tag{4.42}$$

$$\Delta v_{pp}^{H}\Big|^{\max} = \frac{I_{ac}}{4 f_{sw} C_H} \tag{4.43}$$

4.4 Guidelines for Designing the DC-Link Capacitors

The results obtained from the proposed methods for calculating both the switching frequency and the low frequency input voltage ripples for LDN and H-bridge cells in case of single-phase H-bridge inverter with level doubling network can readily be adopted for designing the dc-link capacitors.

A. **LDN cell**

In case of restrictions on the voltage switching ripple, such as limitation of switching noise, conducted EMI, measurements inaccuracy, etc., the dc-link capacitance can be calculated according to (4.42), leading to the same relation obtained in [8] in case of flying capacitor asymmetric H-bridge:

$$C_L \geq \frac{1}{4} \frac{I_{ac}}{f_{sw} \Delta v_{pp}^{L}\Big|^{\max}} \tag{4.44}$$

Instead, if the target is to limit the voltage oscillations at low frequencies (fundamental and/or low-order harmonics), the dc-link capacitance of LDN cell can be calculated by (4.20) or (4.23), leading to:

$$C_L \geq \frac{1}{4} \frac{I_{ac}}{f \; \tilde{v}_{pp}^L}\bigg|^{\max} \tag{4.45}$$

$$C_L \geq \frac{1}{4} \frac{I_{ac}}{f \; V_{1,pp}^L}\bigg|^{\max} \tag{4.46}$$

B. **H-bridge cell**

As for the LDN cell, the dc-link capacitance of the H-bridge cell can be calculated based on the voltage switching ripple according to (4.43), leading to the same relation obtained in [6] in case of basic H-bridge configuration as:

$$C_H \geq \frac{1}{4} \frac{I_{ac}}{f_{sw} \, \Delta v_{pp}^H}\bigg|^{\max} \tag{4.47}$$

Instead, if the maximum dc-link voltage oscillations at the fundamental or the double fundamental frequency is assigned, the dc-link capacitor for the H-bridge cell can be sized by reversing (4.31) and (4.32):

$$C_H \geq \frac{1}{2\pi f \, V_1^H} \frac{[2U_0 - U_2]}{2} I_{ac} \tag{4.48}$$

$$C_H \geq \frac{1}{4\pi f \, V_2^H} \frac{m}{2} I_{ac} \tag{4.49}$$

In practical applications, it could easily be said that low frequency ripple and switching ripple stay in the same ratio as the switching and the fundamental frequencies, that is typically in the range 100–400.

4.5 Simulation and Experimental Results

In order to verify the analytical developments proposed in previous sections, simulations and corresponding experimental tests are carried out for single-phase multilevel H-bridge inverter based on level doubling network. The considered circuit scheme is shown in Fig. 4.11. The main simulation and experimental setup parameters are given in Table 4.1.

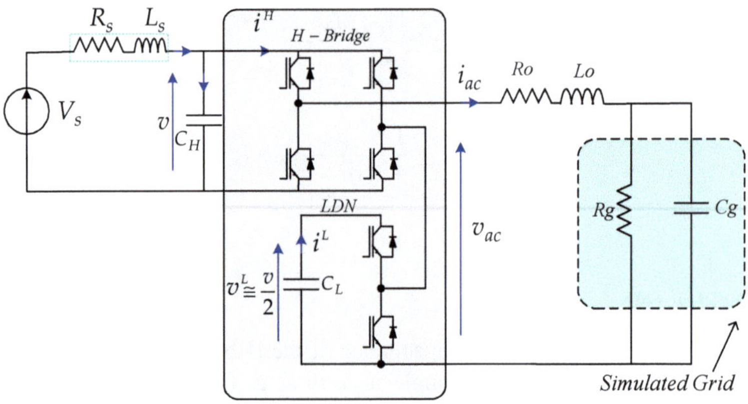

Fig. 4.11 Circuit of the proposed inverter with level doubling network

Table 4.1 Simulation and experimental setup parameters

Parameter	Symbol	Value
DC source voltage	V_s	100 V
DC-link capacitors	C_L, C_H	1.1 mF
Equivalent dc source resistance	R_s	2.5 Ω
Equivalent dc source inductance	L_s	1 μH
Fundamental and switching freq.	f, f_{sw}	50 Hz, 2.5 kHz
Output series impedance	R_o, L_o	6.5 Ω, 34 mH
Equivalent grid (parallel impedance)	R_g, C_g	30 Ω, 44 μF

4.5.1 Simulation Results

Circuit simulations are performed by Matlab/Simulink. In order to verify the matching with the proposed analytical developments, an RLC circuit with unity power factor is used, emulating the grid-connection through an ac-link inductor.

Figure 4.12 presents the alternating LDN input voltage (top diagram), the theoretical low-frequency component $\tilde{v}^L$ calculated by (4.16) and (4.17), corresponding to its value averaged over the switching period (middle diagram), and the switching ripple component together with its theoretical envelopes $\pm\,\Delta v^L{}_{pp}/2$ evaluated from (4.35) and (4.36) (bottom diagram), in case of different modulation indexes to cover the whole modulation range, i.e. for m = 0.25, 0.5, 0.75, and 1.

As for the LDN cell, the alternating H-bridge input voltage (top diagram), its averaged waveform over the switching period (middle diagram), and the

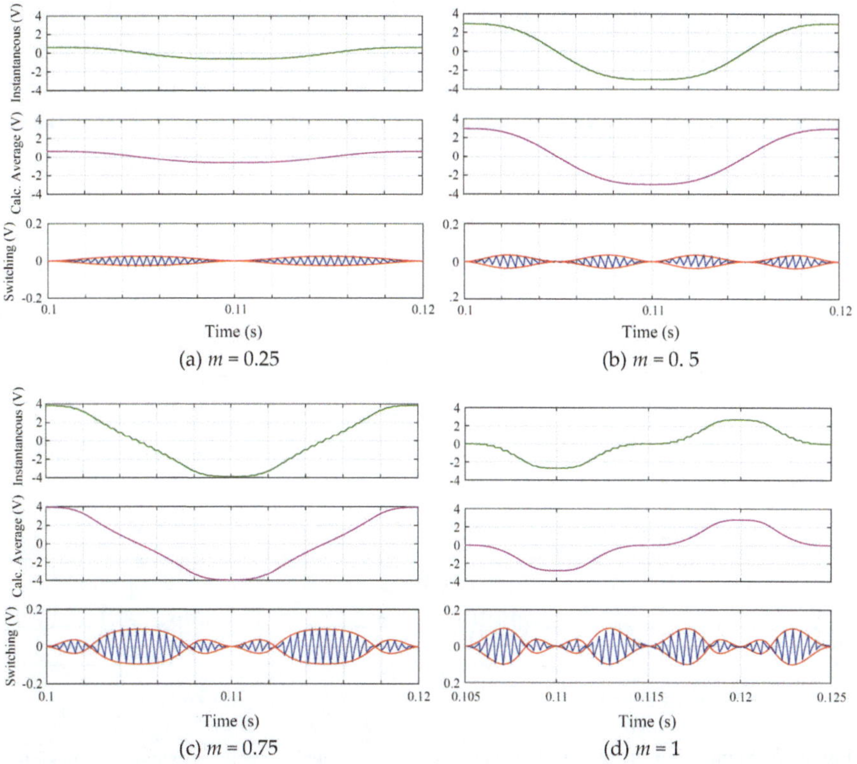

Fig. 4.12 Simulation results: Alternating LDN input voltage (green, top diagrams), theoretical (calculated) low-frequency component (violet, middle diagrams), and switching ripple component with theoretical envelopes (blue and red traces, respectively, bottom diagrams), for different modulation indexes

instantaneous voltage switching ripple Δv^H (blue traces) together with the half peak-to-peak voltage ripple envelope $\pm\Delta v_{pp}^H/2$ (red traces) calculated from (4.39) and (4.40) are depicted in Fig. 4.13 over one fundamental period (T = 20 ms), for m = 0.25, 0.5, 0.75, and 1. A perfect matching is observed for all the considered cases.

4.5.2 *Experimental Results*

Theoretical and simulation results have experimentally been verified by the setup shown in Fig. 4.14. It includes a DSP board (TMS320F28377D) programmed by

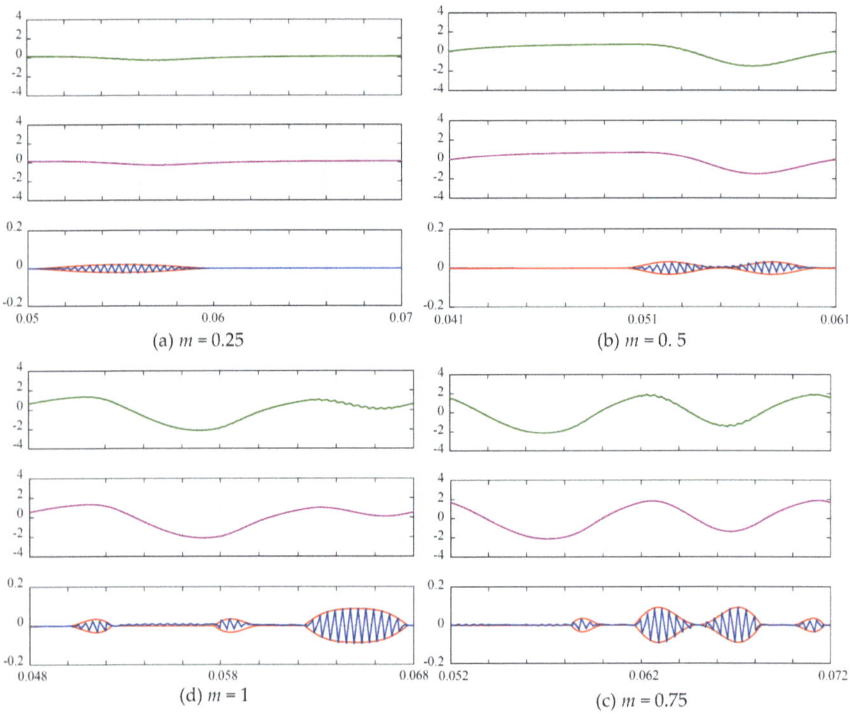

Fig. 4.13 Simulation results: Instantaneous H-bridge input voltage (green, top diagrams), averaged low-frequency component (violet, middle diagrams), and switching ripple component with theoretical envelopes (blue and red traces, respectively, bottom diagrams), for different modulation indexes

Code Composer Studio (CCS), also employed to change online all the modulation parameters. The single-phase H-bridge inverter with level doubling network shown in Fig. 4.11 and has been implemented using two power IGBTs modules IPM PS22A76 (1200 V, 25 A), driven by the DSP board via intermediate optical links. The inverter output is connected to an RLC circuit with unity power factor, emulating the grid-connection through an ac-link inductor.

Experimental results are shown by oscilloscope screenshots, elaborating and emphasizing the signals of interests.

Figures 4.15 and 4.16 show the output variables in case of $m = 0.5$ and $m = 1$. In particular, the top traces represent the total output voltage and current (almost sinusoidal current with unity power factor can be noted), the middle traces represent the individual output voltage of the H-bridge, and the bottom traces represent the individual output voltage of the LDN cell.

Fig. 4.14 View of the experimental setup

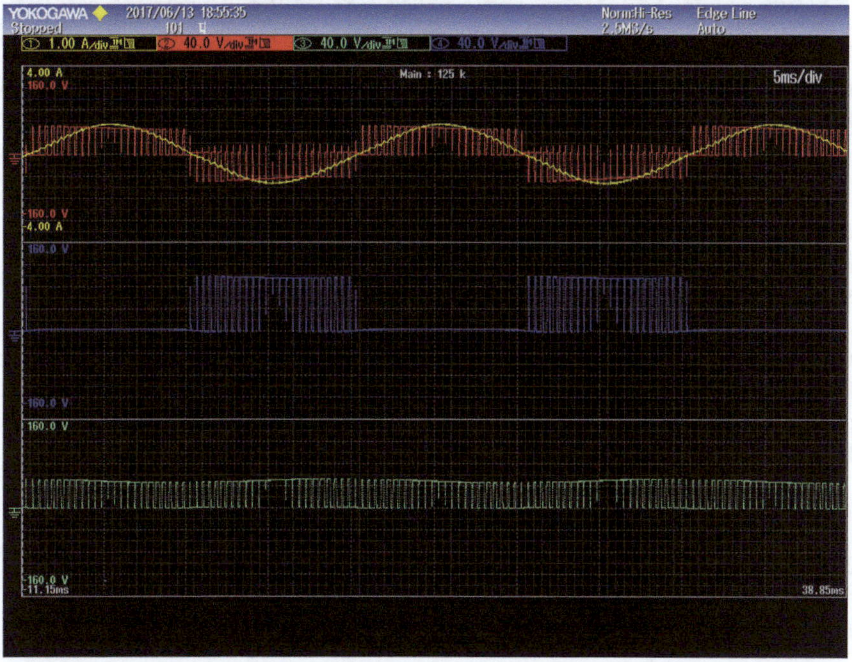

Fig. 4.15 Inverter output variables for $m = 0.5$. Top trace: total voltage and current. Middle trace: H-bridge voltage. Bottom: LDN voltage

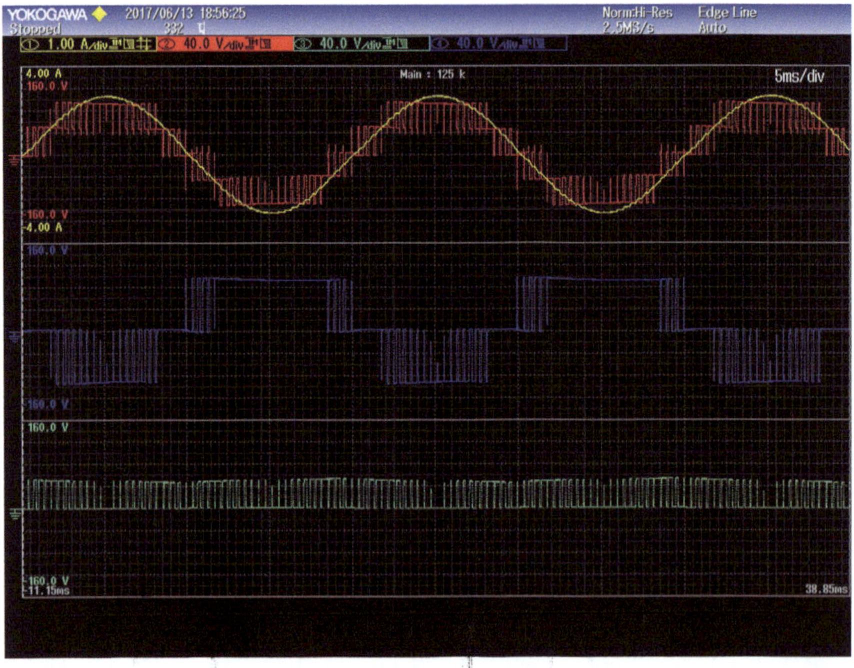

Fig. 4.16 Inverter output variables for $m = 1$. Top trace: total voltage and current. Middle trace: H-bridge voltage. Bottom: LDN voltage

The input voltage of the LDN and H-bridge cells, are shown along with their components in Figs. 4.17 and 4.18, respectively, over one fundamental period ($T = 20$ ms) for different modulation indexes ($m = 0.25$, 0.5, 0.75, and 1), corresponding to the simulation results given in Figs. 4.12 and 4.13.

In particular, the top traces represent the instantaneous voltage, the middle traces represent the low-frequency components, obtained by the built-in low-pass filter function of the oscilloscope, and the bottom traces represent the switching ripple component together with the peak-to-peak envelope analytically calculated on the DSP board and sent to an output DAC channel with a proper voltage scaling.

4.6 Discussion

The alternating LDN and H-bridge input voltages in both simulation and experimental results have been calculated by filtering from the instantaneous voltage its dc component, i.e. V for the H-bridge and $V/2$ for the LDN. In the case of

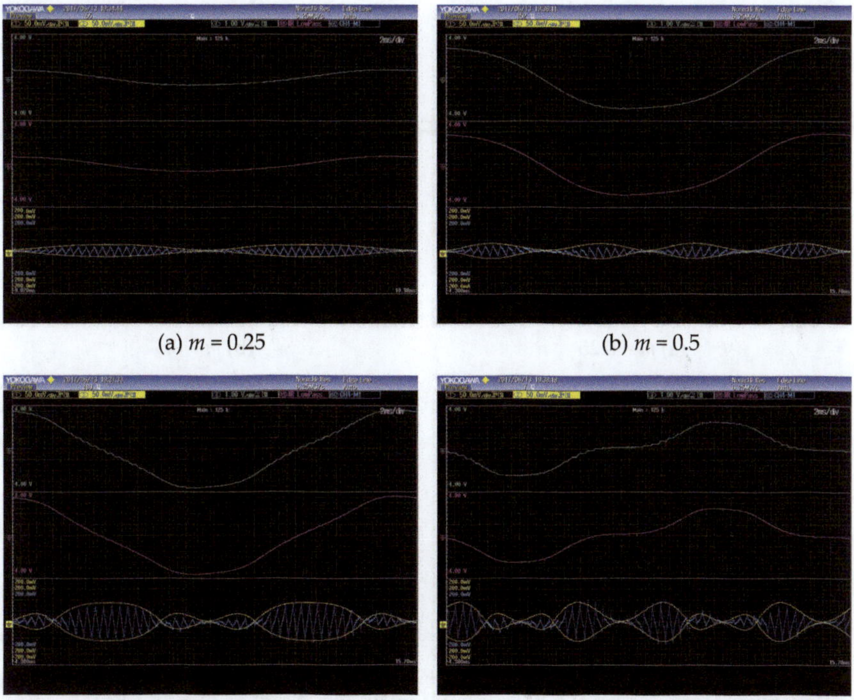

(a) $m = 0.25$ (b) $m = 0.5$

(c) $m = 0.75$ (d) $m = 1$

Fig. 4.17 Experimental results for different modulation indexes. Top traces: alternating LDN input voltage. Middle traces: low-frequency component (filtered by oscilloscope functions). Bottom traces: switching frequency component and corresponding theoretical envelopes. Reference is made to Fig. 4.12

experimental results, the "ac coupling" built-in function of the oscilloscope has been used, introducing a slight and acceptable distortion in the corresponding waveforms.

Simulation and experimental results have a good matching for all the considered cases, as proved by the comparison of Figs. 4.12 and 4.17 for the LDN cell and a comparison of Figs. 4.13 and 4.18 for the H-bridge cell. Note that the same scale is adopted in corresponding diagrams of simulations and experiments to better facilitate the comparison.

The results in terms of peak-to-peak LDN and H-bridge input voltage ripple are summarized in Table 4.2, for the four considered modulation indexes in case of theory (*Th.*), simulations (*Sim.*), and experiments (*Exp.*).

For all the considered cases the matching is within the expected resolution of a few percent.

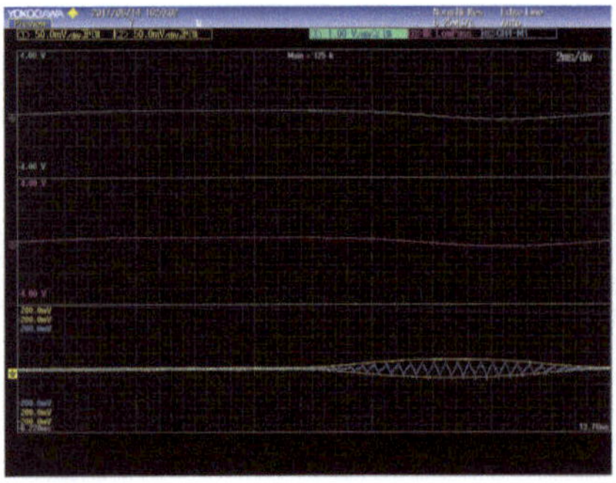

(a) $m = 0.25$

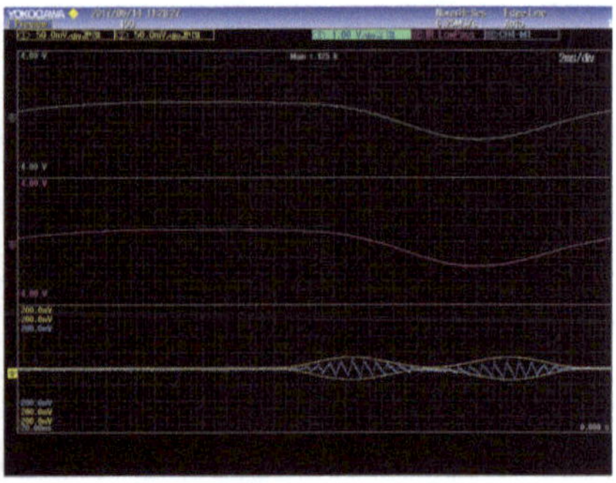

(b) $m = 0.5$

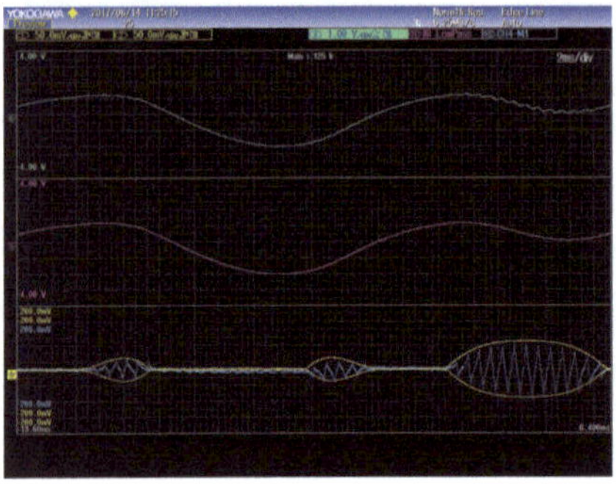

(c) $m = 0.75$

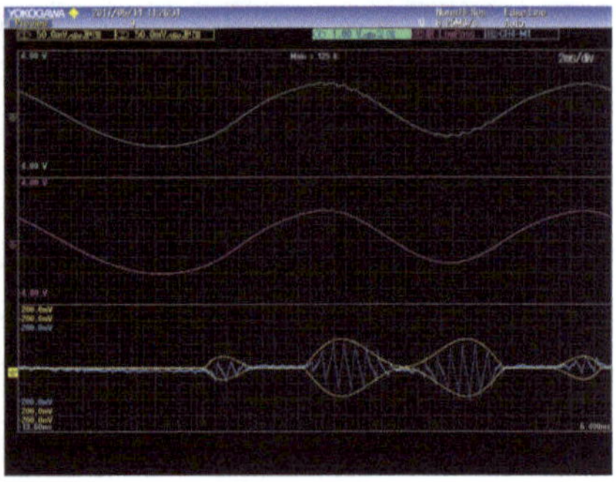

(d) $m = 1$

Fig. 4.18 Experimental results for different modulation indexes. Top traces: alternating H-bridge input voltage. Middle traces: low-frequency component (filtered by oscilloscope functions). Bottom traces: switching frequency component and corresponding theoretical envelopes. Reference is made to Fig. 4.13

Table 4.2 Max peak-to-peak dc-link voltage ripple

m	Switching frequency component of both LDN and H-bridge cells			Low-frequency component of the LDN cell			Low-frequency component of the H-bridge cell	
	Th. (V)	*Sim.* (V)	*Exp.* (V)	*Th.* (V)	*Sim.* (V)	*Exp.* (V)	*Sim.* (V)	*Exp.* (V)
0.25	0.05	0.06	0.06	1.22	1.2	1.2	0.45	0.5
0.50	0.07	0.08	0.08	5.77	5.7	5.6	2.14	2.25
0.75	0.18	0.20	0.20	7.63	7.6	7.4	3.38	3.2
1.00	0.19	0.20	0.20	5.25	5.2	5.0	4	4

References

1. J.W. Kolar, H. Ertl, F.C. Zach, Influence of the modulation method on the conduction and switching losses of a PWM converter system. IEEE Trans. Ind. Appl. **27**(6), 1063–1075 (1991)
2. G. Grandi, J. Loncarski, Evaluation of current ripple amplitude in three-phase PWM voltage source inverters, in *2013 International Conference on Compatibility, Power Electronics*, pp. 156–161 (2013)
3. G. Grandi, J. Loncarski, O. Dordevic, Analysis and comparison of peak-to-peak current ripple in two-level and multilevel PWM inverters. IEEE Trans. Ind. Electron. **62**(5), 2721–2730 (2015)
4. H. Wang, F. Blaabjerg, Reliability of capacitors for DC-link applications in power electronic converters—an overview. IEEE Trans. Ind. Appl. **50**(5), 3569–3578 (2014)
5. M. Vujacic, M. Srndovic, M. Hammami, G. Grandi, Evaluation of DC voltage ripple in single-phase H-bridge PWM inverters, in *IECON Proceedings (Industrial Electronics Conference)*, pp. 3235–3240 (2016)
6. M. Vujacic, M. Hammami, M. Srndovic, G. Grandi, Theoretical and experimental investigation of switching ripple in the DC-link voltage of single-phase H-bridge PWM inverters. Energies **10**(8), 1–16 (2017)
7. G. Buticchi, E. Lorenzani, G. Franceschini, A five-level single-phase grid-connected converter for renewable distributed systems. IEEE Trans. Ind. Electron. **60**(3), 906–918 (2013)
8. Y. Zhang, L. Sun, An efficient control strategy for a five-level inverter comprising flying-capacitor asymmetric H-bridge. IEEE Trans. Ind. Electron. **58**(9), 4000–4009 (2011)
9. S.K. Chattopadhyay, C. Chakraborty, A new multilevel inverter topology with self-balancing level doubling network. IEEE Trans. Ind. Electron. **61**(9), 4622–4631 (2014)
10. S.K. Chattopadhyay, C. Chakraborty, A new asymmetric multilevel inverter topology suitable for solar PV applications with varying irradiance. IEEE Trans. Sustain. Energy **3029**(99), 1 (2017)

Authored Papers

11. M. Vujacic, M. Srndovic, M. Hammami, G. Grandi, Evaluation of DC voltage ripple in single-phase H-bridge PWM inverters, in *42nd Annual Conference Industrial Electronics Society, IECON 2016*, 24–27 October 2016, Florence, Italy
12. M. Vujacic, M. Hammami, M. Srndovic, G. Grandi, Theoretical and experimental investigation of DC voltage switching ripple in single-phase H-bridge PWM inverters. Energies **10**(8), art. 1189 (2017)
13. M. Hammami, M. Vujacic, A. Viatkin, G. Grandi, Analysis of a flexible single-phase multilevel inverter topology for photovoltaic applications, in *9th International Renewable Energy Congress, IREC 2018*, 20–22 March 2018, Hammamet, Tunisia

Chapter 5
Analysis of DC-Link Current and Voltage Ripple: Three-Phase Configuration

5.1 Introduction

Three-phase inverters are customarily adopted, due to their wide availability, in many different applications such as variable speed ac drives, uninterruptable power supplies (UPS), stand-alone and grid-connected systems, etc. With reference to the three-phase voltage source inverters (VSIs), many papers investigated the inverter output characteristics (voltage and current). However, not many of them focused on the inverter input (dc-link) side. The peak-to-peak output current ripple amplitude has been calculated over the fundamental period, and presented together with numerical and experimental verification in the case of two-level and multilevel VSIs, in [1, 2]. A comparison of the output current ripple in case of multiphase inverters is presented in [3], considering different number of phases.

A complete analysis of dc-link voltage ripple for three-phase two-level inverters is presented in [4], considering both switching frequency and double fundamental frequency components. The developments related to input current and voltage characteristics are usually based on the Fourier analysis (harmonics) and rms calculations. The analysis and minimization of the input current and voltage ripple rms, and the comparison under several reference signals is given in [5]. In [6] the dc-link current ripple characteristics in three-phase inverters with balanced load have been investigated basing on the Fourier analysis.

A detailed analysis of the dc-link capacitor current in three-phase three-level neutral-point-clamped (NPC) and cascaded H-bridge (CHB) inverters, which provides the basis for dc-link capacitor sizing in these topologies, have been done in [7]. In particular, the rms value and low frequency harmonics of the capacitor current have been calculated numerically only in case of sinusoidal PWM (SPWM). With reference to the three-phase three-level CHB, the analysis of dc-link voltage ripple and the design of dc-link capacitor in case of the centered PWM (CPWM) have not been reported in the literature yet.

M. Hammami, *Level Doubling Network and Ripple Correlation Control MPPT Algorithm for Grid-Connected Photovoltaic Systems*, Springer Theses,
https://doi.org/10.1007/978-3-030-10492-4_5

The three-phase level doubling network (LDN) topology has been presented in [8, 9] and the concept of self balancing capability of the capacitor was analyzed in case of nearest voltage level control (staircase modulation). Nowadays, the analysis of dc-link voltage ripple in the case of three-phase LDN configuration for both sinusoidal and centered PWMs has not been reported in the literature yet.

A complete analysis of the dc-link voltage ripple for three-phase H-bridge configuration as well as for three-phase LDN configuration, considering both switching frequency and double fundamental frequency components will be presented in this chapter. Simple and effective guidelines to design the dc-link capacitor are given on the basis of the dc-voltage ripple requirements for different modulation strategies.

5.2 Analysis of DC-Link Ripple: Basic H-Bridge Configuration

A preliminary theoretical analysis of the dc-link voltage ripple, considering both switching frequency and double-fundamental frequency components, has been presented in [10], for the single-phase H-bridge inverter. The typical configuration of three-phase three-level HB inverter is shown in Fig. 5.1. It consists of dc voltage source (V_s) with series inductive filter (L_s) and equivalent resistance (R_s) and a dc-link capacitor (C_H).

In the case SPWM, the calculations for single-phase H-bridge inverter, developed in [10, 11], can be applied directly to the three-phase configuration. The calculation of input current and voltage ripples, and the design of the capacitor for each phase are exactly the same as for the single-phase configuration.

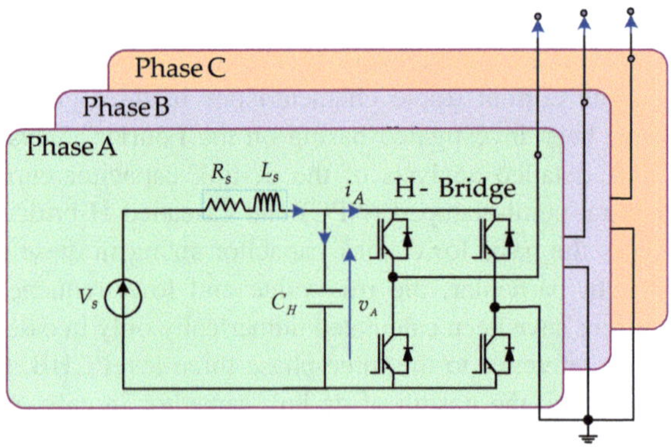

Fig. 5.1 Basic circuit scheme of the three-phase H-bridge inverter

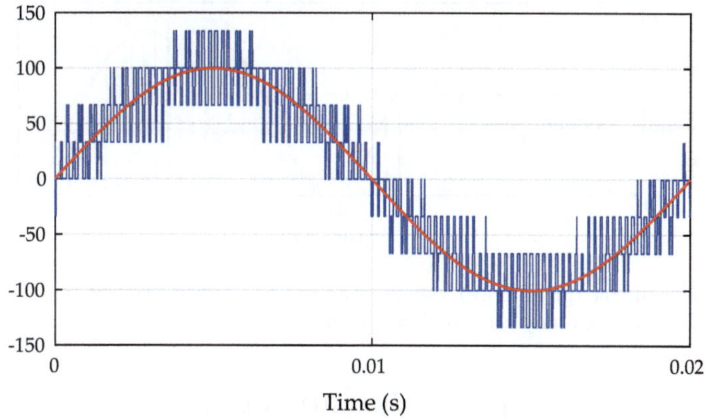

Fig. 5.2 Ideal PWM inverter output voltage and its averaged counterpart in case of constant dc source $V = 100$ V and $m = 1$

In order to maximize the linear modulation index range and/or reduce the output current ripple, a common-mode can be injected into the individual inverter voltages. In this case, the modulating signals normalized by the dc component of the H-bridge dc-link voltage V, are expressed as:

$$\begin{cases} u_{ac}^{A} = m\sin(\vartheta) + c_m(\vartheta) \\ u_{ac}^{B} = m\sin\left(\vartheta - \frac{2}{3}\pi\right) + c_m(\vartheta) \\ u_{ac}^{C} = m\sin\left(\vartheta + \frac{2}{3}\pi\right) + c_m(\vartheta) \end{cases} \tag{5.1}$$

The common-mode injection is explained in details in Chap. 3. The centred PWM (min-max) is approximated as a triangular waveform as:

$$c_m \cong \frac{2m}{\pi^2}\left[\sin(3\vartheta) - \frac{1}{9}\sin(9\vartheta)\right] \tag{5.2}$$

The output voltage of phase A and its fundamental component (averaged value) are presented in Fig. 5.2.

5.2.1 Input Current Analysis

Figure 5.3 shows the instantaneous inverter input current $i_A(t)$ of phase A in case of CPWM. The input current has three relevant components: dc (average) component I, low frequency component $\tilde{i}(t)$, and switching frequency component Δi(t):

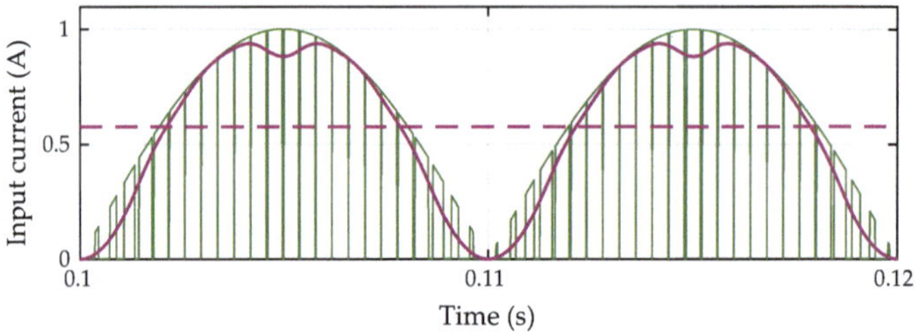

Fig. 5.3 Input current (green trace) and its averaged counterpart (violent traces), in case of $m = 1.15$ and $I_{ac} = 1$ A (unity sinusoidal output current)

$$\begin{cases} i_A(t) = I_A + \tilde{i}_A(t) + \Delta i_A(t) \\ i_B(t) = I_B + \tilde{i}_B(t) + \Delta i_B(t) \\ i_C(t) = I_C + \tilde{i}_C(t) + \Delta i_C(t) \end{cases} \tag{5.3}$$

With reference to phase A, the averaged input current can be calculated on the basis of the inverter input/output power balance as:

$$\bar{i}_A = u_{ac}^A \, i_{ac}^A \tag{5.4}$$

being i_{ac}^A the output current of phase A expressed as:

$$i_{ac}^A = I_{ac} \, \sin(\vartheta) \tag{5.5}$$

where I_{ac} is the output current amplitude. The same results can be obtained for phases B and C considering the displacement of 120°.

Introducing (5.1), (5.2) and (5.5) in (5.4), the averaged input current of phase A in case of CPWM is expressed as:

$$\bar{i}_A = I_{ac} \sin\vartheta \left[m \sin\vartheta + \frac{2\,m}{\pi^2} \sin(3\vartheta) - \frac{2\,m}{9\,\pi^2} \sin(9\vartheta) \right] \tag{5.6}$$

resulting in:

$$\begin{aligned} \bar{i}_A = & \frac{m\,I_{ac}}{2} + \frac{m\,I_{ac}}{2} \left(\frac{2}{\pi^2} - 1 \right) \cos(2\vartheta) + \frac{m\,I_{ac}}{\pi^2} \cos(4\vartheta) \\ & - \frac{m\,I_{ac}}{9\pi^2} \left(\cos(8\vartheta) - \cos(10\vartheta) \right) \end{aligned} \tag{5.7}$$

Equation (5.7) can also be expressed as:

$$\bar{i}_A = I_A + \sum_k I_{k,A} \cos(k\vartheta) \tag{5.8}$$

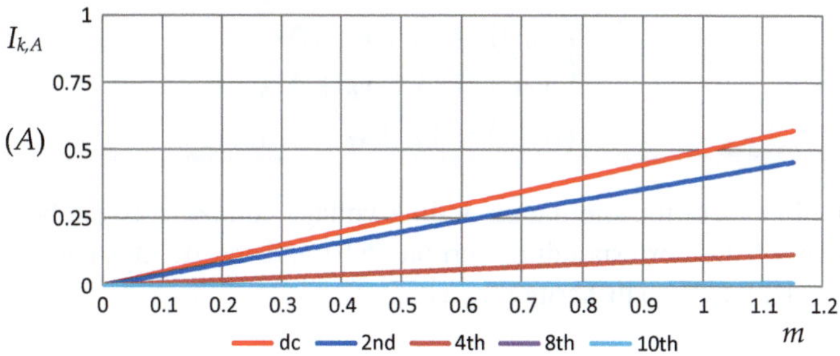

Fig. 5.4 Amplitudes of low order harmonic of input current as a function of m (I_{ac} = 1 A)

being I_A the average (dc) current, and $I_{k,A}$ the amplitude of the kth harmonic, which can readily be obtained from (5.7) and (5.8):

$$\begin{cases} I_A = \dfrac{m\,I_{ac}}{2} \\ I_{2,A} = \dfrac{m\,I_{ac}}{2}\left(1 - \dfrac{2}{\pi^2}\right) \\ I_{4,A} = \dfrac{m\,I_{ac}}{\pi^2} \\ I_{8,A} = I_{10,A} = \dfrac{m\,I_{ac}}{9\pi^2} \end{cases} \tag{5.9}$$

The amplitudes of the low order harmonics input current are depicted in Fig. 5.4 as a function of the modulation index m.

Once the averaged input current of phase A is calculated from (5.6), the switching frequency input current component is finally calculated from (5.3) as:

$$\begin{aligned} \Delta i_A &= i_A - \bar{i}_A \\ &= I_{ac} \sin\vartheta \left(1 - m\sin\vartheta + \frac{2\,m}{\pi^2}\sin(3\vartheta) - \frac{2\,m}{9\,\pi^2}\sin(9\vartheta) + \frac{2\,m}{25\,\pi^2}\sin(15\vartheta)\right) \end{aligned} \tag{5.10}$$

5.2.2 *DC-Link Voltage Analysis*

The instantaneous dc-link voltage consists of three relevant components: dc (average) component V, alternating (low-frequency) components $\tilde{v}$, and switching frequency component Δv:

$$\begin{cases} v_A(t) = V_A + \tilde{v}_A + \Delta v_A \\ v_B(t) = V_B + \tilde{v}_B + \Delta v_B \\ v_C(t) = V_C + \tilde{v}_C + \Delta v_C \end{cases} \tag{5.11}$$

The amplitudes of the alternating dc-link voltage components can be determined on the basis of the corresponding current (5.9) and the dc-link equivalent impedance Z_k, calculated at kth harmonic, as

$$V_{k,A} = I_{k,A}\, Z_k \tag{5.12}$$

being Z_k the parallel between the reactance of the dc-link capacitor C_H, and the equivalent dc source resistance R_S, both calculated for the kth harmonic as:

$$Z_k = \frac{R_s}{\sqrt{k^2\, R_s^2\, \omega^2\, C_H^2 + 1}} \tag{5.13}$$

Considering (5.12), (5.13) and (5.9), the amplitudes of the alternating dc-link voltage components at double fundamental frequency (100 Hz) in case of CPWM can be calculated as:

$$V_{2,A} = \frac{m\, I_{ac}}{2}\left(1 - \frac{2}{\pi^2}\right) Z_2 \tag{5.14}$$

In the case of SPWM, the amplitude of the second order harmonic is the same as for single-phase H-bridge configuration, expressed as:

$$V_{2,A} = \frac{m\, I_{ac}}{2}\, Z_2 \tag{5.15}$$

It can be noticed that the second order harmonic (100 Hz) voltage oscillations decreases in case of CPWM (5.14) compared to the case of SPWM (5.15). The dc-link capacitor can be sized by using (5.14) or (5.15) in case of CPWM or SPWM, respectively.

5.3 Analysis of DC-Link Ripple: H-Bridge Plus LDN Configuration

The three-phase LDN configuration depicted in Fig. 5.5.

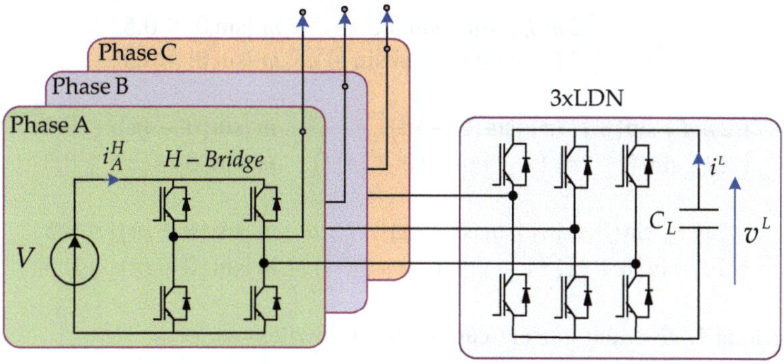

Fig. 5.5 The three-phase H-bridge LDN configuration with a single dc-link

5.3.1 Analysis of the Input Currents

A. **LDN cell**

As for the single-phase, the instantaneous LDN input currents i_A^L, i_B^L and i_C^L can be written as:

$$\begin{cases} i_A^L = \bar{i}_A^L + \Delta i_A^L = I_A^L + \tilde{i}_A^L + \Delta i_A^L \\ i_B^L = \bar{i}_B^L + \Delta i_B^L = I_B^L + \tilde{i}_B^L + \Delta i_B^L \\ i_C^L = \bar{i}_C^L + \Delta i_C^L = I_C^L + \tilde{i}_C^L + \Delta i_C^L \end{cases} \tag{5.16}$$

Being $\bar{i}_A^L$ the averaged value over the switching period, consisting in dc component I_A^L, $\tilde{i}_A^L$ the low-frequency harmonic components, and Δi_A^L the instantaneous switching ripple of phase A and similarly for phases B and C.

With reference to sinusoidal output currents with unity power factor, the instantaneous output currents are:

$$\begin{cases} i_{ac}^A = I_{ac}\sin(\vartheta) \\ i_{ac}^B = I_{ac}\sin(\vartheta - \frac{2}{3}\pi) \\ i_{ac}^C = I_{ac}\sin(\vartheta + \frac{2}{3}\pi) \end{cases} \tag{5.17}$$

where I_{ac} is the output current amplitude.

On the basis of the input-output power balance of the LDN and considering the averaged quantities over the switching period, the input LDN current has been analytically developed in case of single-phase inverter (see Chap. 4). The same expression can be used directly for phase A, B and C to find:

$$\tilde{i}_A^L = \begin{cases} 2\,m\,I_{ac}\,\sin\vartheta|\sin\vartheta|, & m\,|\sin\vartheta| \leq 0.5 \\ 2\,I_{ac}\,\sin\vartheta\,(1 - m|\sin\vartheta|), & m\,|\sin\vartheta| \geq 0.5 \end{cases} \tag{5.18}$$

$$\tilde{i}_B^L = \begin{cases} 2\,m\,I_{ac}\sin\left(\vartheta - \frac{2}{3}\pi\right)\left|\sin\left(\vartheta - \frac{2}{3}\pi\right)\right|, & m\left|\sin\left(\vartheta - \frac{2}{3}\pi\right)\right| \leq 0.5 \\ 2\,I_{ac}\sin\left(\vartheta - \frac{2}{3}\pi\right)\left(1 - m\left|\sin\left(\vartheta - \frac{2}{3}\pi\right)\right|\right), & m\left|\sin\left(\vartheta - \frac{2}{3}\pi\right)\right| \geq 0.5 \end{cases} \tag{5.19}$$

$$\tilde{i}_C^L = \begin{cases} 2\,m\,I_{ac}\sin\left(\vartheta + \frac{2}{3}\pi\right)\left|\sin\left(\vartheta + \frac{2}{3}\pi\right)\right|, & m\left|\sin\left(\vartheta + \frac{2}{3}\pi\right)\right| \leq 0.5 \\ 2\,I_{ac}\sin\left(\vartheta + \frac{2}{3}\pi\right)\left(1 - m\left|\sin\left(\vartheta + \frac{2}{3}\pi\right)\right|\right), & m\left|\sin\left(\vartheta + \frac{2}{3}\pi\right)\right| \geq 0.5 \end{cases} \tag{5.20}$$

The total LDN input current can simply be written as:

$$\tilde{i}^L = \tilde{i}_A^L + \tilde{i}_B^L + \tilde{i}_C^L \tag{5.21}$$

The amplitudes of the low-frequency LDN input current components have been calculated analytically for the single-phase configuration as explained in Chap. 4:

$$\begin{cases} I^L = 0 \\ I_1^L = (2U_0 - U_2)\,I_{ac} \\ I_2^L = 0 \\ I_n^L = |U_{n+1} - U_{n-1}|\,I_{ac} \end{cases} \quad (n \geq 3 \text{ odd integer}) \tag{5.22}$$

where U_0, U_2 and U_n are defined in the Appendix B.

Considering (5.22) and due to the modulation symmetry among the three phases (120° displacement), all odd harmonics except those that are a multiple of 3 are eliminated, while the harmonics multiples of 3 are tripled, this resulting in:

$$\begin{cases} I_h^L = 0 \\ I_{3k}^L = 3|U_{3k+1} - U_{3k-1}|I_{ac} \end{cases} \quad (h, k \,\text{odd integer},\; h \neq 3k) \tag{5.23}$$

Figure 5.6 presents the amplitudes of the low-frequency LDN input current harmonics in case of SPWM, as a function of the modulation index, considering (5.22) and (5.23) with unity output current, I_{ac} = 1 A.

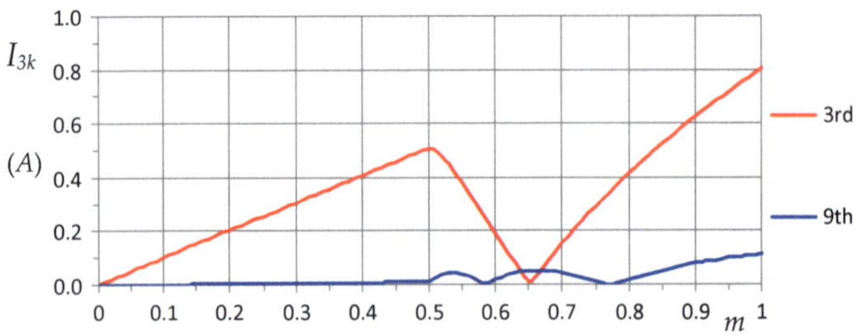

Fig. 5.6 Amplitudes of low-frequency LDN input current harmonics as a function of the modulation index in case of SPWM (I_{ac} = 1A)

The total LDN input current $\tilde{i}^L$, its low-frequency component calculated from (5.18), (5.19), (5.20) and (5.21), corresponding to the averaged value over the switching period, and its dc component (which is equal to zero) are shown in Fig. 5.7.

In case of CPWM, the normalized modulating signals are expressed in (5.1). Replacing (5.1) in (5.18), (5.19) and (5.20), the input LDN currents for phase A, B and C becomes:

$$\tilde{i}_A^L = \begin{cases} 2\,I_{ac}\,\sin\vartheta\left|m\sin(\vartheta)+c_m(\vartheta)\right|, & \left|m\sin(\vartheta)+c_m(\vartheta)\right|\le 0.577 \\ 2\,I_{ac}\,\sin\vartheta\left(1-\left|m\sin(\vartheta)+c_m(\vartheta)\right|\right), & \left|m\sin(\vartheta)+c_m(\vartheta)\right|\ge 0.577 \end{cases} \tag{5.24}$$

$$\tilde{i}_B^L = \begin{cases} 2\,I_{ac}\sin\left(\vartheta-\tfrac{2}{3}\pi\right)\left|m\sin\left(\vartheta-\tfrac{2}{3}\pi\right)+c_m(\vartheta)\right|, & \left|m\sin\left(\vartheta-\tfrac{2}{3}\pi\right)+c_m(\vartheta)\right|\le 0.577 \\ 2\,I_{ac}\sin\left(\vartheta-\tfrac{2}{3}\pi\right)\left(1-\left|m\sin\left(\vartheta-\tfrac{2}{3}\pi\right)+c_m(\vartheta)\right|\right), & \left|m\sin\left(\vartheta-\tfrac{2}{3}\pi\right)+c_m(\vartheta)\right|\ge 0.577 \end{cases} \tag{5.25}$$

$$\tilde{i}_C^L = \begin{cases} 2\,I_{ac}\sin\left(\vartheta+\tfrac{2}{3}\pi\right)\left|m\sin\left(\vartheta+\tfrac{2}{3}\pi\right)+c_m(\vartheta)\right|, & \left|m\sin\left(\vartheta+\tfrac{2}{3}\pi\right)+c_m(\vartheta)\right|\le 0.577 \\ 2\,I_{ac}\sin\left(\vartheta+\tfrac{2}{3}\pi\right)\left(1-\left|m\sin\left(\vartheta+\tfrac{2}{3}\pi\right)+c_m(\vartheta)\right|\right), & \left|m\sin\left(\vartheta+\tfrac{2}{3}\pi\right)+c_m(\vartheta)\right|\ge 0.577 \end{cases} \tag{5.26}$$

The low frequency components of the total input current (5.21) can easily be calculated by adding (5.24), (5.25) and (5.26). In this case, the total LDN input current is depicted in Fig. 5.8 for the two relevant cases m = 0.577 and m = 1.15, corresponding to the new modulation boundaries in case of centered PWM.

It can be noticed from Figs. 5.7 and 5.8 that CPWM mitigates the LDN low-frequency harmonics. It has also been verified that harmonics can further be reduced by exploiting the degree of freedom offered by modifying the common-mode signal (mCPWM). It has been found empirically that the simple constant gain of 1.15 applied to the common-mode centering signal (mCPWM) further reduces the LDN low-frequency current ripple $\tilde{i}^L$ in the whole modulation range, as shown in Fig. 5.9.

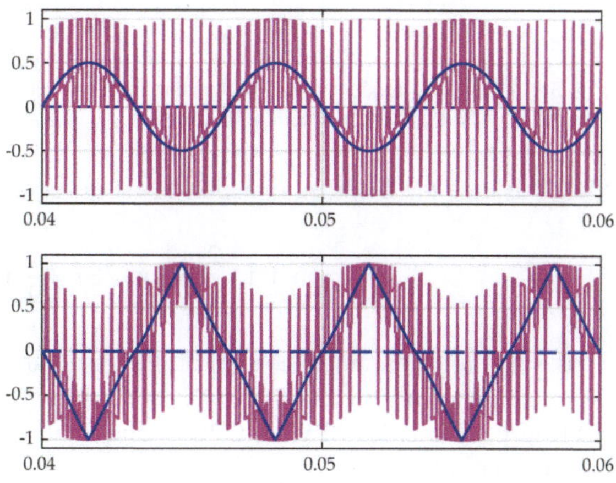

Fig. 5.7 Total LDN input current (pink traces) and its averaged counterpart (bleu traces) for m = 0.5 (top) and m = 1 (bottom) in case of unity output current (I_{ac} = 1 A)

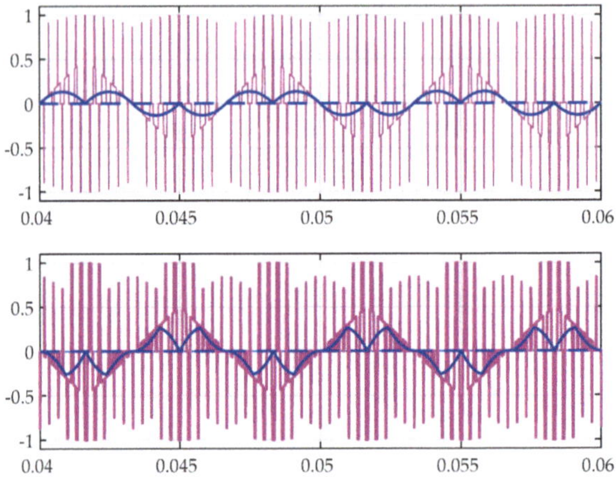

Fig. 5.8 Total LDN input current (pink traces) and its averaged counterpart (bleu traces) for $m = 0.577$ (top) and $m = 1.15$ (bottom) in case of unity output current ($I_{ac} = 1$ A) for CPWM

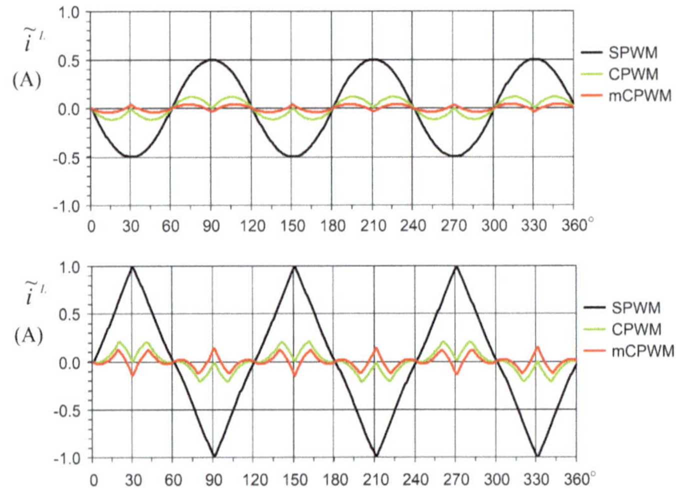

Fig. 5.9 LDN low-frequency dc-link current ripple in case of SPWM, CPWM, and modified CPWM for $m = 0.5$ (top) and $m = 1$ (bottom), ($I_{ac} = 1$ A)

In this case the modified common-mode injection (mCPWM) can be written as:

$$c_m \cong 1.15\,\frac{2m}{\pi^2}\left[\sin(3\vartheta) - \frac{1}{9}\sin(9\vartheta)\right] \tag{5.27}$$

Basing on the harmonic content of the input current given by (5.22) and (5.23). The amplitudes of 3th and 9th LDN input current are depicted in Figs. 5.10 and 5.11, respectively, for SPWM, CPWM and modified CPWM as a function of the modulation index.

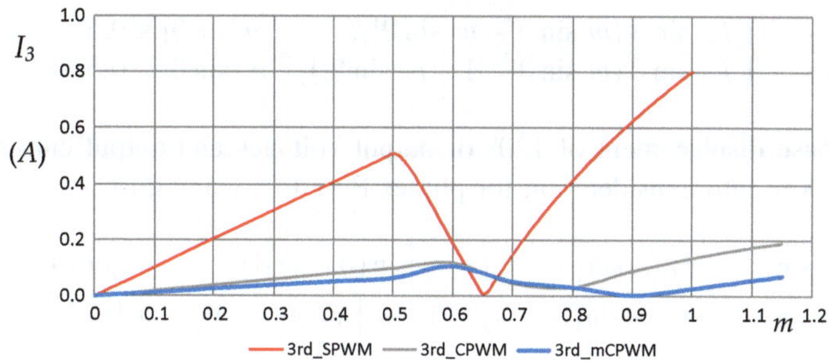

Fig. 5.10 Third order harmonic of LDN input current in case of SPWM, CPWM, and modified CPWM (I_{ac} = 1 A)

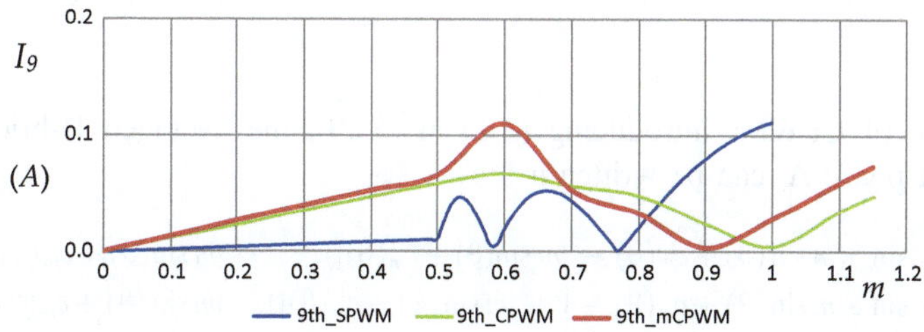

Fig. 5.11 Ninth order harmonic of LDN input current in case of SPWM, CPWM, and modified CPWM (I_{ac} = 1A)

B. **H-bridge cell**

The instantaneous H-bridge input currents for phase A, B and C consist of three relevant components: the dc (average) components I_A^H, I_B^H and I_C^H, the low frequency components $\tilde{i}_A^H$,$\tilde{i}_B^H$ and $\tilde{i}_C^H$, and the switching frequency component Δi_A^H, Δi_B^H and Δi_C^H, respectively as:

$$\begin{cases} i_A^H = \bar{i}_A^H + \Delta i_A^H = I_A^H + \tilde{i}_A^H + \Delta i_A^H \\ i_B^H = \bar{i}_B^H + \Delta i_B^H = I_B^H + \tilde{i}_B^H + \Delta i_B^H \\ i_C^H = \bar{i}_C^H + \Delta i_C^H = I_C^H + \tilde{i}_C^H + \Delta i_C^H \end{cases} \tag{5.28}$$

Based on the analysis of the H-bridge input current in case of single-phase configuration, simple and straightforward calculations can be applied in case of three-phase configuration (Fig. 5.5) considering both SPWM and CPWM.

In case of sinusoidal modulation (SPWM), the analytical expression of the low frequency components of the H-bridge input current, that has been calculated in case of single-phase configuration, can directly be applied to phase A of the three-phase configuration, since each H-bridge cell is supplied with an independent dc source, leading to:

$$\bar{i}_A^H = \begin{cases} I_{ac} \sin\vartheta\,(m\sin\vartheta - m\,|\sin\vartheta|), & m\,|\sin\vartheta| \le 0.5 \\ I_{ac} \sin\vartheta\,(m\sin\vartheta - 1 + m\,|\sin\vartheta|), & m\,|\sin\vartheta| \ge 0.5 \end{cases} \tag{5.29}$$

The phase displacement of 120° of output voltages and output currents (5.17) must be taken into consideration for phases B and C, leading to:

$$\bar{i}_B^H = \begin{cases} I_{ac}\sin\left(\vartheta - \frac{2}{3}\pi\right)\left(m\sin\left(\vartheta - \frac{2}{3}\pi\right) - m\left|\sin\left(\vartheta - \frac{2}{3}\pi\right)\right|\right), & m\left|\sin\left(\vartheta - \frac{2}{3}\pi\right)\right| \le 0.5 \\ I_{ac}\sin\left(\vartheta - \frac{2}{3}\pi\right)\left(m\sin\left(\vartheta - \frac{2}{3}\pi\right) - 1 + m\left|\sin\left(\vartheta - \frac{2}{3}\pi\right)\right|\right), & m\left|\sin\left(\vartheta - \frac{2}{3}\pi\right)\right| \ge 0.5 \end{cases} \tag{5.30}$$

$$\bar{i}_C^H = \begin{cases} I_{ac}\sin\left(\vartheta + \frac{2}{3}\pi\right)\left(m\sin\left(\vartheta + \frac{2}{3}\pi\right) - m\left|\sin\left(\vartheta + \frac{2}{3}\pi\right)\right|\right), & m\left|\sin\left(\vartheta + \frac{2}{3}\pi\right)\right| \le 0.5 \\ I_{ac}\sin\left(\vartheta + \frac{2}{3}\pi\right)\left(m\sin\left(\vartheta + \frac{2}{3}\pi\right) - 1 + m\left|\sin\left(\vartheta + \frac{2}{3}\pi\right)\right|\right), & m\left|\sin\left(\vartheta + \frac{2}{3}\pi\right)\right| \ge 0.5 \end{cases} \tag{5.31}$$

In case of CPWM, introducing (5.1) in (5.29), the averaged H-bridge input current of phase A, can be written as:

$$\bar{i}_A^H = \begin{cases} I_{ac}\,\sin\vartheta(m\sin(\vartheta) + c_m(\vartheta) - \,|m\sin(\vartheta) + c_m(\vartheta)|), & |m\sin(\vartheta) + c_m(\vartheta)| \le 0.577 \\ I_{ac}\,\sin\vartheta(m\sin(\vartheta) + c_m(\vartheta) - 1 + \,|m\sin(\vartheta) + c_m(\vartheta)|), & |m\sin(\vartheta) + c_m(\vartheta)| \ge 0.577 \end{cases} \tag{5.32}$$

The same holds for phase B and C considering the phase displacement and introducing (5.1) in (5.30) and (5.31), respectively.

Figure 5.12 shows the instantaneous H-bridge input current of phase A, its low-frequency component calculated from (5.32), corresponding to the averaged value over the switching period, and its dc component in case of CPWM.

The amplitudes of the H-bridge input current in case of CPWM (min-max) have been calculated numerically and shown in Fig. 5.13.

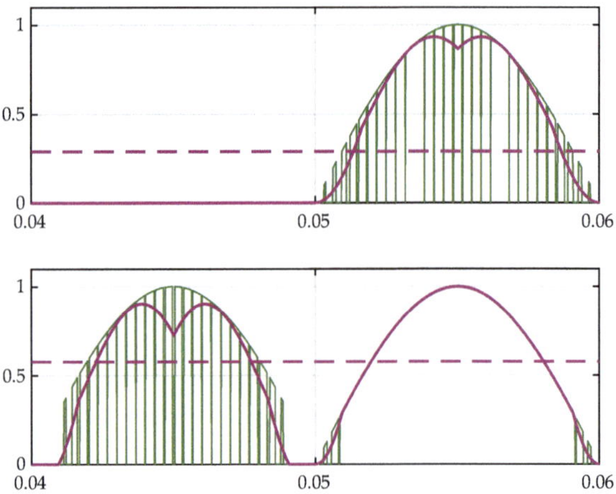

Fig. 5.12 H-bridge input current (green traces) and its averaged counterpart (violet traces) for m = 0.577 (top) and m = 1.15 (bottom) in case of unity output current (I_{ac} = 1 A)

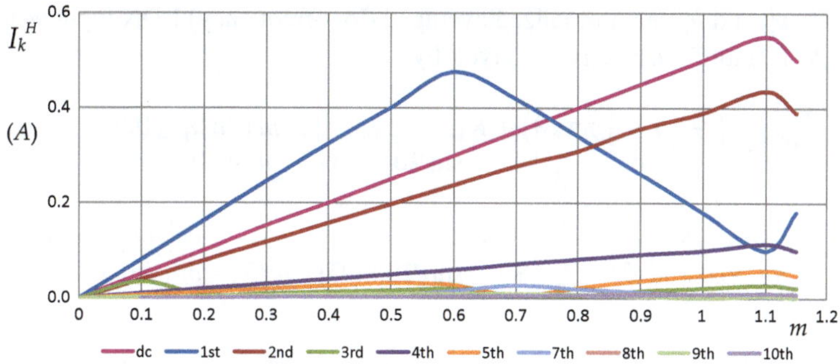

Fig. 5.13 Amplitudes of H-bridge input current in case of CPWM (I_{ac} = 1 A)

5.3.2 Analysis of Input Voltages

In order to design the LDN and H-bridge capacitors properly, analytical analyses of dc-link voltages are performed. To insure a satisfactory design of capacitors, the peak-to-peak values of the voltage oscillations must be known.

As for the input currents, the instantaneous dc-link voltage for both LDN and H-bridge cells can be written as:

$$v = V + \tilde{v} + \Delta v \tag{5.33}$$

being V the average input voltage, $\tilde{v}$ the low frequency components and Δv the switching frequency component.

A. **LDN cell**

The low-frequency dc-link voltage component of the LDN cell can be calculated analytically by integrating the total LDN input current (5.21) as:

$$\tilde{v}^L = \frac{1}{\omega C_L} \int \tilde{i}^L(\vartheta)\, d\vartheta + K = \frac{1}{\omega C_L} \int \left(\tilde{i}_A^L + \tilde{i}_B^L + \tilde{i}_C^L\right) d\vartheta + K \tag{5.34}$$

where K is an integration constant.

In the case of SPWM, introducing (5.18), (5.19) and (5.20) in (5.34) and solving the integral, the input voltage of each phase A, B and C can be written as:

$$\begin{cases} \tilde{v}_A^L = \dfrac{I_{ac}}{2fC_L}\, \tilde{u}_A^L \\ \tilde{v}_B^L = \dfrac{I_{ac}}{2f\, C_L}\, \tilde{u}_B^L \\ \tilde{v}_C^L = \dfrac{I_{ac}}{2fC_L}\, \tilde{u}_C^L \end{cases} \tag{5.35}$$

being $\tilde{u}_A^L$, $\tilde{u}_B^L$ and $\tilde{u}_C^L$ the normalized averaged (low-frequency) LDN input voltage of phase A, B and C, respectively, given by

$$\tilde{u}_A^L = \begin{cases} \frac{m}{\pi}\left(\vartheta - \frac{1}{2}\sin 2\vartheta\right) + K_1, & m\left|\sin\vartheta\right| \leq 0.5 \\ -\frac{2}{\pi}\cos\vartheta - \frac{m}{\pi}\left(\vartheta - \frac{1}{2}\sin 2\vartheta\right) + K_2, & m\left|\sin\vartheta\right| \geq 0.5 \end{cases} \tag{5.36}$$

$$\tilde{u}_B^L = \begin{cases} \frac{m}{\pi}\left(\vartheta - \frac{1}{2}\sin\left(2\vartheta - \frac{4\pi}{3}\right)\right) + K_1, & m\left|\sin\left(\vartheta - \frac{2}{3}\pi\right)\right| \leq 0.5 \\ -\frac{2}{\pi}\cos\left(\vartheta - \frac{2\pi}{3}\right) - \frac{m}{\pi}\left(\vartheta - \frac{1}{2}\sin\left(2\vartheta - \frac{4\pi}{3}\right)\right) + K_2, & m\left|\sin\left(\vartheta - \frac{2}{3}\pi\right)\right| \geq 0.5 \end{cases} \tag{5.37}$$

$$\tilde{u}_C^L = \begin{cases} \frac{m}{\pi}\left(\vartheta - \frac{1}{2}\sin\left(2\vartheta + \frac{4\pi}{3}\right)\right) + K_1, & m\left|\sin\left(\vartheta + \frac{2}{3}\pi\right)\right| \leq 0.5 \\ -\frac{2}{\pi}\cos\left(\vartheta + \frac{2\pi}{3}\right) - \frac{m}{\pi}\left(\vartheta - \frac{1}{2}\sin\left(2\vartheta + \frac{4\pi}{3}\right)\right) + K_2, & m\left|\sin\left(\vartheta + \frac{2}{3}\pi\right)\right| \geq 0.5 \end{cases} \tag{5.38}$$

The values of K_1 and K_2 can be determined by setting the continuity of the voltage between the different regions. Being $\tilde{u}^L$ the sum of the three normalized averaged $\tilde{u}_A^L$, $\tilde{u}_B^L$ and $\tilde{u}_C^L$ of the LDN input voltage of phase A, B and C, respectively, the waveform of $\tilde{u}^L$ is depicted in Fig. 5.14 in the whole fundamental period, for different modulation indexes.

Instead of integrating the total LDN input current to get the corresponding input voltage, the amplitudes of the kth voltage harmonic can be calculated multiplying the amplitude of the corresponding current harmonic (5.23) by the kth order reactance $1/(k\omega C_L)$ as:

$$\tilde{v}_{3k}^L = \frac{I_{3k}^L}{k\,\omega\,C_L} = \frac{3I_{ac}\left|U_{3k+1} - U_{3k-1}\right|}{2\pi\,kfC_L} \tag{5.39}$$

Equation (5.39) suggests the following normalization for $\tilde{v}_{3k}^L$:

$$\tilde{v}_{3k}^L = \frac{I_{ac}}{2fC_L}\tilde{u}_{3k}^L \tag{5.40}$$

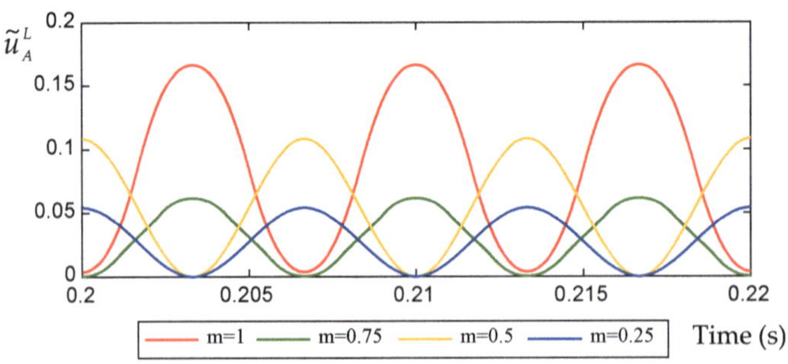

Fig. 5.14 Normalized low-frequency LDN input voltage of phase A in fundamental period, in case of SPWM, for different modulation indices, m = 0.25, 0.5, 0.75 and 1

being $\tilde{u}_{3k}^{L}$ the normalized amplitude of the LDN low-frequency voltage expressed as:

$$\tilde{u}_{3k}^{L} = \frac{3|U_{3k+1} - U_{3k-1}|}{k\pi} \tag{5.41}$$

Considering the third order harmonic (highest amplitude) and utilizing (5.41), the normalized peak-to-peak LDN voltage ripple is written as:

$$\tilde{u}_{3,pp}^{L} = \frac{2|U_4 - U_2|}{\pi} \tag{5.42}$$

B. **H-bridge cell**

In three-phase LDN configuration (Fig. 5.5), each H-bridge is supplied by an independent dc source, which means that the analytical developments of the H-bridge input voltage components (in the single-phase) are exactly the same for phase A considering the sinusoidal PWM. The H-bridge input voltage of phase B and C will be displaced by 120°. In this case, the voltage amplitudes at first and second harmonic have been calculated as (see Chap. 4):

$$V_1^H = \frac{1}{\omega C_H}\frac{[2U_0 - U_2]}{2}I_{ac} = \frac{1}{2\pi f C_H}\frac{[2U_0 - U_2]}{2}I_{ac} \tag{5.43}$$

$$V_2^H = \frac{1}{2\,\omega C_H}\frac{m}{2}I_{ac} = \frac{1}{4\pi f C_H}\frac{m}{2}I_{ac} \tag{5.44}$$

The amplitudes of the input voltage of the H-bridge cell can be calculated numerically based on the input current for the H-bridge cell at the *k*th harmonic order for both CPWM and mCPWM.

5.3.3 Guidelines for Designing the DC-Link Capacitor

A. **LDN cell**

The normalized peak-to-peak LDN input voltage $\tilde{u}_{3,pp}^{L}$ has been calculated analytically from (5.41) in case of sinusoidal PWM, However, due to the cumbersome calculations for both CPWM and modified CPWM, it has been calculated numerically. Figure 5.15 shows the normalized peak-to-peak LDN input voltage as a function of the modulation index in case of SPWM, CPWM and mCPWM.

The maximum peak-to-peak LDN voltage ripple can be determined according to Fig. 5.15 for SPWM, CPWM and mCPWM respectively as:

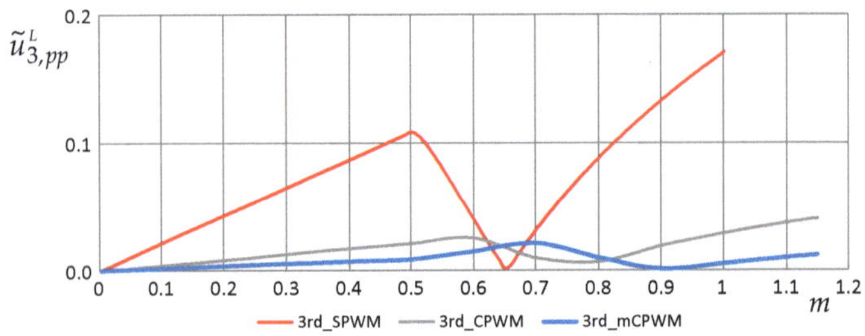

Fig. 5.15 Normalized peak-to-peak LDN input voltage as a function of modulation index in case of SPWM, CPWM and mCPWM

$$\tilde{u}^L_{3,pp}\Big|^{\max} = 0.169 \cong \frac{1}{6} \quad \text{for SPWM } (m = 1) \tag{5.45}$$

$$\tilde{u}^L_{3,pp}\Big|^{\max} = 0.04 = \frac{1}{25} \quad \text{for CPWM } (m = 1.15) \tag{5.46}$$

$$\tilde{u}^L_{3,pp}\Big|^{\max} = 0.02 = \frac{1}{50} \quad \text{for mCPWM } (m = 0.7) \tag{5.47}$$

Finally, using (5.42), (5.45), (5.46) and (5.47), the maximum amplitude of the peak-to-peak voltage ripple for different modulations can be written as:

$$\tilde{v}^L_{3,pp}\Big|^{\max} \cong \frac{I_{ac}}{12f\,C_L} \quad \text{for SPWM} \tag{5.48}$$

$$\tilde{v}^L_{3,pp}\Big|^{\max} \cong \frac{I_{ac}}{50f\,C_L} \quad \text{for CPWM} \tag{5.49}$$

$$\tilde{v}^L_{3,pp}\Big|^{\max} \cong \frac{I_{ac}}{100f\,C_L} \quad \text{for mCPWM} \tag{5.50}$$

The dc-link LDN capacitance can be calculated according to (5.48), (5.49) and (5.50), for SPWM, CPWM and mCPWM, respectively, leading to

$$C_L \geq \frac{I_{ac}}{12f\tilde{v}^L_{3,pp}\Big|^{\max}} \quad \text{for SPWM} \tag{5.51}$$

$$C_L \geq \frac{I_{ac}}{50f\tilde{v}^L_{3,pp}\Big|^{\max}} \quad \text{for CPWM} \tag{5.52}$$

$$C_L \geq \frac{I_{ac}}{100 f \tilde{v}_{3,pp}^L \Big|^{\max}} \quad \text{for mCPWM} \tag{5.53}$$

It can be noticed that the modulation strategy has an influence on the dimensioning of the capacitor of the LDN cell.

B. **H-bridge cell**

For the H-bridge cell, the dc-link capacitance can be calculated by (5.34) and (5.44), leading to:

$$C_H \geq \frac{1}{2\pi f V_1^H} \frac{[2U_0 - U_2]}{2} I_{ac} \tag{5.54}$$

$$C_H \geq \frac{1}{4\pi f V_2^H} \frac{m}{2} I_{ac} \tag{5.55}$$

5.3.4 Simulation Results

In order to verify the results obtained from the analytical developments, Matlab/Simulink simulations have been carried out for the three-phase H-bridge LDN configuration (Fig. 5.5) controlled by the SPWM, CPWM and modified CPWM. The inverter output is connected to a three-phase RLC circuit with unity power factor, emulating the grid-connection through an ac-link inductor as shown in Fig. 5.16.

Table 5.1 summarizes the circuit parameters and components used in the simulations.

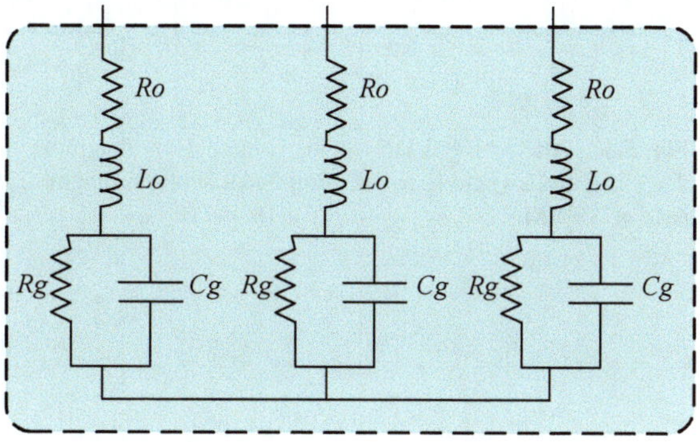

Fig. 5.16 Electrical circuit scheme of the RLC load

Table 5.1 Simulation setup parameters

Parameter	Symbol	Value
DC source voltage	V	100 V
DC-link LDN capacitor	C_L	1.1 mF
Fundamental and switching frequencies	f, f_{sw}	50 Hz, 2.5 kHz
Output series impedance	R_o, L_o	6.5 Ω, 34 mH
Equivalent grid (parallel impedance)	R_g, C_g	30 Ω, 44 μF

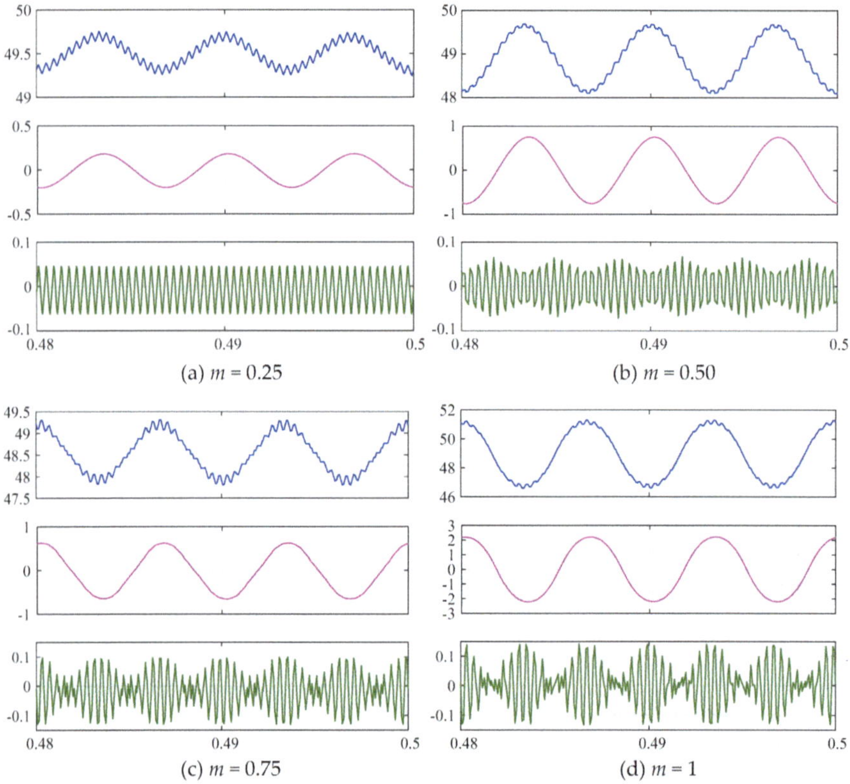

Fig. 5.17 Alternating LDN input voltage (top diagram), low-frequency component (middle diagram), and switching ripple component (bottom diagram), for different modulation indexes in case of SPWM

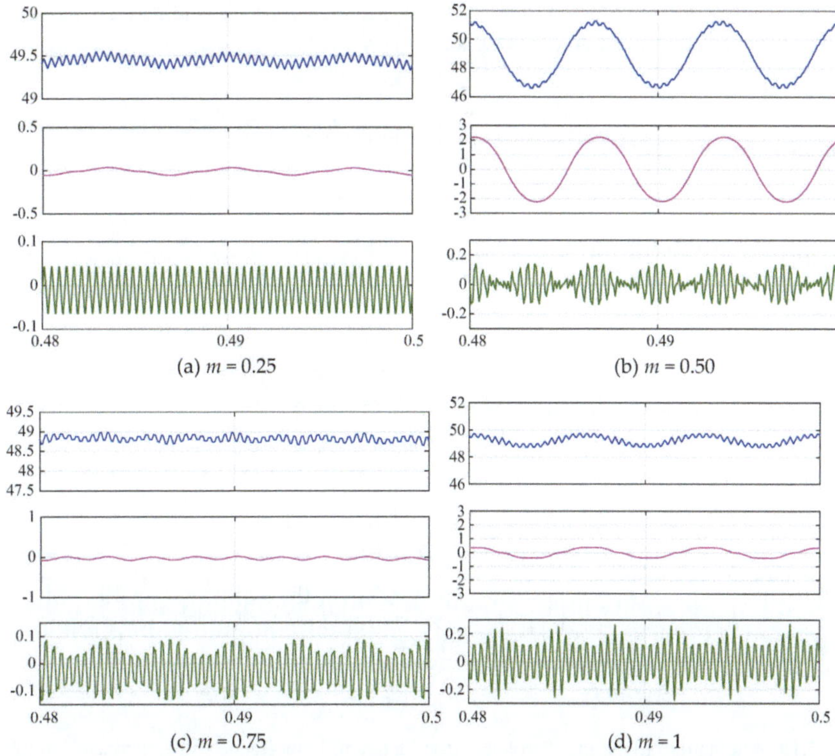

Fig. 5.18 Alternating LDN input voltage (top diagram), low-frequency component (middle diagram), and switching ripple component (bottom diagram), for different modulation indexes in case of CPWM

The instantaneous dc-link LDN voltage with its components are shown in Fig. 5.17 for SPWM, in Fig. 5.18 for CPWM, and in Fig. 5.19 for mCPWM over one fundamental period ($T = 20$ ms) and for different modulation indexes ($m = 0.25$, 0.5, 0.75, and 1). In particular, the top diagram represents the instantaneous voltage, the middle diagram represents the low-frequency components, calculated analytically, and the bottom diagram represents the switching ripple component.

5.4 Discussion

The theoretical (*Th.*) and simulations (*Sim.*) results in terms of peak-to-peak LDN input voltage ripple are summarized in Table 5.2 for the four considered modulation indexes in case of SPWM, CPWM and mCPWM.

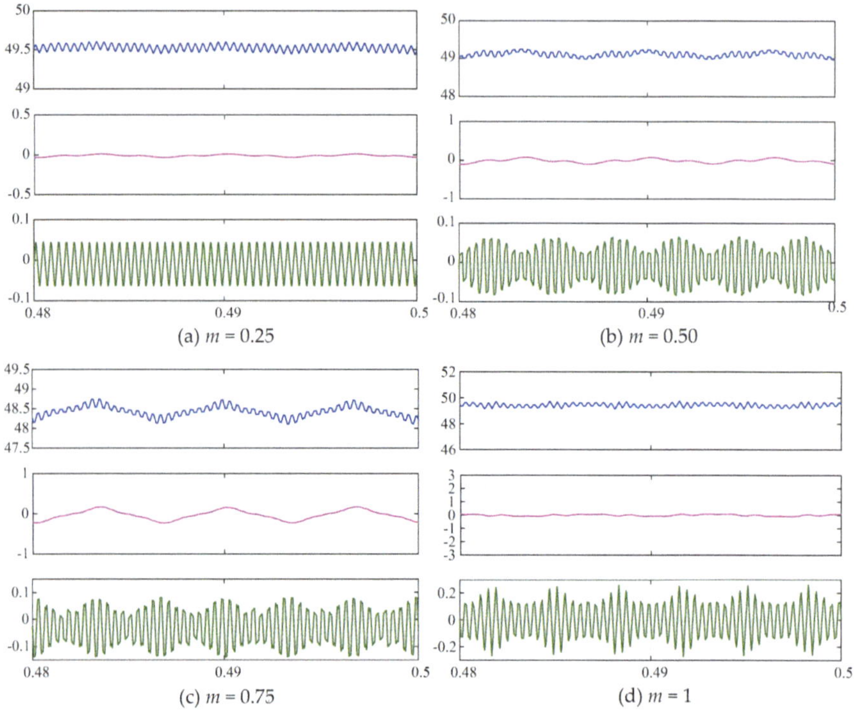

Fig. 5.19 Alternating LDN input voltage (top diagram), low-frequency component (middle diagram), and switching ripple component (bottom diagram), for different modulation indexes in case of mCPWM

Table 5.2 Max peak-to-peak dc-link voltage ripple of the LDN cell

m	Low-frequency component—SPWM		Low-frequency component—CPWM		Low-frequency component—mCPWM	
	Th. (V)	*Sim.* (V)	*Th.* (V)	*Sim.* (V)	*Th.* (V)	*Sim.* (V)
0.25	0.38	0.38	0.080	0.088	0.028	0.026
0.50	1.5	1.5	0.29	0.30	0.12	0.15
0.75	1.3	1.3	0.081	0.089	0.51	0.48
1.00	4.5	4.4	0.78	0.73	0.14	0.14

For all cases considered, the matching is within the expected resolution of a few percent. It can also be noticed that, by changing the modulation strategy, there is a noticeable decrease in the low frequency dc-link LDN voltage ripple, this leading to more relaxed constraints in the design of the dc-link LDN capacitor.

References

1. G. Grandi, J. Loncarski, Evaluation of current ripple amplitude in three-phase PWM voltage source inverters, in *2013 International Conference on Compatibility and Power Electronics* (2013), pp. 156–161
2. G. Grandi, J. Loncarski, O. Dordevic, Analysis and comparison of peak-to-peak current ripple in two-level and multilevel PWM inverters. IEEE Trans. Ind. Electron. **62**(5), 2721–2730 (2015)
3. G. Grandi, J. Loncarski, C. Rossi, Comparison of peak-to-peak current ripple amplitude in multiphase PWM voltage source inverters, in *2013 15th European Conference on Power Electronics and Applications, EPE 2013* (2013), pp. 1–10
4. M. Vujacic, M. Hammami, M. Srndovic, G. Grandi, Evaluation of DC voltage ripple in single-phase H-bridge PWM inverters, in *IEEE 26th International Symposium on Industrial Electronics (ISIE)* (2017), pp. 711–716
5. P.A. Dahono, Y. Sato, T. Kataoka, Analysis and minimization of ripple components of input current and voltage of PWM inverters. IEEE Trans. Ind. Appl. **32**(4), 945–950 (1996)
6. M.H. Bierhoff, F.W. Fuchs, DC-link harmonics of three-phase voltage-source converters influenced by the pulsewidth-modulation strategy—an analysis. IEEE Trans. Ind. Electron. **55** (5), 2085–2110 (2008)
7. G.I. Orfanoudakis, S.M. Sharkh, M.A. Yuratich, Analysis of dc-link capacitor current in three-level neutral point clamped and cascaded H-bridge inverters. IET Power Electron. **6**(7), 1376–1389 (2013)
8. S.K. Chattopadhyay, C. Chakraborty, A new multilevel inverter topology with self-balancing level doubling network. IEEE Trans. Ind. Electron. **61**(9), 4622–4631 (2014)
9. S.K. Chattopadhyay, C. Chakraborty, A new asymmetric multilevel inverter topology suitable for solar PV applications with varying irradiance. IEEE Trans. Sustain. Energy **3029**(99), 1 (2017)
10. M. Vujacic, M. Srndovic, M. Hammami, G. Grandi, Evaluation of DC voltage ripple in single-phase H-bridge PWM inverters, in *IECON 2016—42nd Annual Conference of the IEEE Industrial Electronics Society* (2016), pp. 3235–3240
11. M. Vujacic, M. Hammami, M. Srndovic, G. Grandi, Theoretical and experimental investigation of switching ripple in the DC-link voltage of single-phase H-bridge PWM inverters. Energies **10**(8), 1–16 (2017)

Authored Papers

12. M. Vujacic, M. Hammami, M. Srndovic, G. Grandi, Evaluation of DC voltage Ripple in three-phase PWM Voltage Source Inverters, in *26th Institute of Electrical and Electronics Engineers (IEEE) International Symposium on Industrial Electronics, ISIE 2017* (2017)
13. M. Vujacic, M. Srndovic, M. Hammami, G. Grandi, Evaluation of DC voltage ripple in single-phase H-bridge PWM inverters, in *42nd Annual Conference of the IEEE Industrial Electronics Society, IECON 2016*, Florence, Italy, 24–27 Oct 2016
14. M. Vujacic, M. Hammami, M. Srndovic, G. Grandi, Theoretical and experimental investigation of DC voltage switching ripple in single-phase H-bridge PWM inverters. Energies **10**(8), 1189 (2017)

Chapter 6
Improvement of RCC-MPPT in Case of Multiple DC-Link Voltage Harmonics

6.1 Introduction

The basic ripple correlation control (RCC) maximum power point tracking (MPPT) algorithm has been proposed in [1–3]. Similar methods are presented in [4, 5] in which the moving average (*MAvg*) concept has been used instead of high/low-pass filters to identify the ripple signals (i.e. 2nd harmonics, 100 Hz components). A modified RCC-MPPT algorithm has been introduced in [6]. In order to smooth out the instability introduced by fast irradiance transients, an hybrid RCC-MPPT method is introduced in [5]. However, the literature does not provide analysis of the RCC algorithm in case of multiple PV voltage and current harmonics, as in the case of multilevel inverters. The implementation of MPPT schemes by RCC algorithm is discussed in case of multiple ripple harmonics. A single-phase single-stage grid-connected photovoltaic (PV) system based on H-bridge inverter and level doubling network (LDN) is considered, leading to a multilevel inverter having the double of output voltage levels compared to the basic H-bridge inverter topology. The LDN is basically a half-bridge fed by a floating capacitor, with voltage self-balancing capability. In this case, additional voltage and current harmonics are introduced in the input dc-link of the H-bridge, leading to multiple PV voltage and current low-order harmonics comparing to the 2nd harmonic components in case of basic H-bridge configuration. Voltage and current harmonic components are used in RCC schemes to estimate the voltage derivative of the power dP/dV and to drive the PV operating point to the MPP. Different possible dP/dV estimations have been proposed in case of multiple harmonics. The steady-state and transient performances of the proposed RCC-MPPT schemes have been numerically tested and compared by MATLAB/Simulink. Results have been verified by experimental tests considering the whole multilevel PV generation system.

M. Hammami, *Level Doubling Network and Ripple Correlation Control MPPT Algorithm for Grid-Connected Photovoltaic Systems*, Springer Theses,
https://doi.org/10.1007/978-3-030-10492-4_6

6.2 PV Voltage and Current Analysis

In the considered single-phase single-stage PV generation system (Fig. 6.1), the PV field is directly connected to the input side of the H-bridge, so the PV field voltage corresponds to the H-bridge input dc-link voltage, whereas the PV current can be calculated on the basis of H-bridge input dc-link current and the whole dc-link impedance, as shown in the following.

6.2.1 H-Bridge Input Current Analysis

The instantaneous input dc-link current i^H of the H-bridge is composed by its averaged value over the switching period $\bar{i}^H$, and the instantaneous switching ripple Δi^H. Introducing also the average over the fundamental period I^H (dc component) and the alternating low-frequency harmonic component $\tilde{i}^H$, it becomes:

$$i^H = \bar{i}^H + \Delta i^H = I^H + \tilde{i}^H + \Delta i^H \tag{6.1}$$

The analysis of the H-bridge input current has been given in details in Chap. 4. All the components have been calculated analytically. In fact, the H-bridge input current averaged over the switching period is given by (6.2).

$$\bar{i}^H = I^H + \tilde{i}^H = \begin{cases} I_{ac} \sin\vartheta\,(m \sin\vartheta - m\,|\sin\vartheta|), & m|\sin\vartheta| \leq 0.5 \\ I_{ac} \sin\vartheta(m \sin\vartheta - 1 + m|\sin\vartheta|), & m|\sin\vartheta| \geq 0.5 \end{cases} \tag{6.2}$$

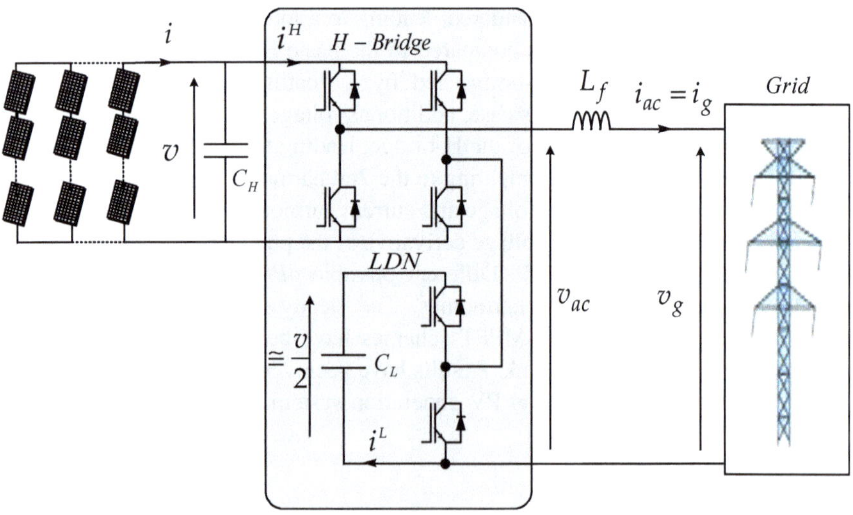

Fig. 6.1 Block diagram of the single-phase LDN multilevel inverter

Furthermore, the harmonic spectrum of the averaged H-bridge input current has been calculated analytically in Chap. 4 as:

$$\bar{i}^H = \frac{1}{2}\left[m + (2U_0 - U_2)\sin\vartheta - m\cos 2\vartheta - \sum_{n=3}^{\infty}(U_{n+1} - U_{n-1})\sin n\vartheta\right] I_{ac} \quad (6.3)$$

where $n \geq 3$ is odd and I_{ac} is the amplitude of output current.

6.2.2 PV Voltage Analysis

As for the current in Eq. (6.1), the instantaneous H-bridge dc-link voltage v, corresponding to the PV voltage in the considered single-stage PV generation scheme, is composed by its averaged value over the switching period $\bar{v}$ and the instantaneous switching ripple Δv. Introducing the average over the fundamental period V (dc component) and the alternating low-frequency harmonic component $\tilde{v}$, it becomes:

$$v = \bar{v} + \Delta v = V + \tilde{v} + \Delta v \cong V + \tilde{v} \quad (6.4)$$

The switching frequency component Δv is strongly smoothed by the dc-link capacitor, and it is usually negligible for switching frequencies starting from few kHz. It is assumed here $\Delta v \cong 0$.

The amplitude of the kth order harmonic component V_k has been calculated in Chap. 4 as:

$$V_k = Z_k I_k^H \quad (6.5)$$

being Z_k the total dc-link impedance for the kth harmonic, given by the parallel between the reactance of the dc-link capacitor (C_H), and the equivalent resistance of the PV field ($R_{PV} \cong V_{MPP}/I_{MPP}$ in the vicinity of the MPP):

$$Z_k = \frac{R_{PV}}{\sqrt{(R_{PV} k\omega C_H)^2 + 1}} \quad (6.6)$$

Considering realistic parameter values for PV fields and dc-link capacitors, the assumption $R_{PV} \gg 1/(k\omega C_H)$ is generally satisfied, and (6.6) becomes:

$$Z_k = \frac{1}{k\omega C_H} \quad (6.7)$$

Using the previous assumption, (6.5) can be written as:

$$V_k = \frac{1}{k\omega C_H} I_k^H \tag{6.8}$$

Considering only the 1st and 2nd dominating harmonic components, the amplitude of individual harmonic is needed in the proposed RCC schemes to estimate the voltage derivative of the power *dP*/*dV* and to drive the PV operating point to the MPP.

In addition to the calculation of the amplitude of individual PV voltage harmonics, the peak-to-peak of the PV voltage oscillation is evaluated, being the amplitude of this oscillation generally lower than the sum of all the harmonic amplitudes.

As known, the equivalent PV series resistance R_{PV} around the MPP (i.e., V_{MPP}/I_{MPP}) is generally much higher than the dc-link capacitor reactance $1/(k\omega C_H)$, considering realistic parameter values for PV fields and dc-link capacitors C_H. As a consequence, the most of the alternating input current component of the H-bridge is circulating on the dc-link capacitor, and the alternating dc-link (PV) voltage component can be calculated by integrating the alternating current component $\tilde{i}^H$ given by (6.2), being $I^H = mI_{ac}/2$, as expressed in (6.3).

In this case, the peak-to-peak PV voltage low-frequency alternating component can be calculated as:

$$\tilde{v}_{pp} = \begin{cases} \frac{I_{ac}}{\omega C_H} \int\limits_{\pi+\arcsin\left(\frac{m}{2}\right)}^{2\pi-\arcsin\left(\frac{m}{2}\right)} \left[\frac{m}{2} - m\cos(2\vartheta)\right] \mathrm{d}\vartheta, & m \le 0.5 \\ \frac{I_{ac}}{\omega C_H} \left\{ \int\limits_{\pi+\arcsin\left(\frac{1}{2m}\right)}^{2\pi-\arcsin\left(\frac{1}{2m}\right)} \left[-\frac{m}{2} - \sin(\vartheta)\right] \mathrm{d}\vartheta + 2 \int\limits_{\arcsin\left(\frac{m}{2}\right)}^{\arcsin\left(\frac{1}{2m}\right)} \left[\frac{m}{2} - m\cos(2\vartheta)\right] \mathrm{d}\vartheta \right\}, & m \ge 0.5 \end{cases} \tag{6.9}$$

Equation (6.9) can be written as:

$$\tilde{v}_{pp} = \frac{I_{ac}}{f\, C_H}\, \tilde{u}_{pp} \tag{6.10}$$

being $\tilde{u}_{pp}$ the normalized low-frequency PV voltage component, given by:

$$\tilde{u}_{pp} = \begin{cases} \frac{1}{2\pi}\left[-m\arcsin\left(\frac{m}{2}\right) + \frac{m\pi}{2} + \frac{m^2}{2}\sqrt{4-m^2}\right], & m \le 0.5 \\ \frac{1}{2\pi}\left[2m\arcsin\left(\frac{1}{2m}\right) - \frac{m\pi}{2} + \sqrt{1-\frac{1}{4m^2}} - m\arcsin\left(\frac{m}{2}\right) + \frac{m^2}{2}\sqrt{4-m^2}\right], & m \ge 0.5 \end{cases} \tag{6.11}$$

Figure 6.2 shows the normalized peak-to-peak PV voltage low-frequency alternating component as a function of modulation index, in case of unity sinusoidal output current with unity power factor, for H-bridge configuration (green dashed

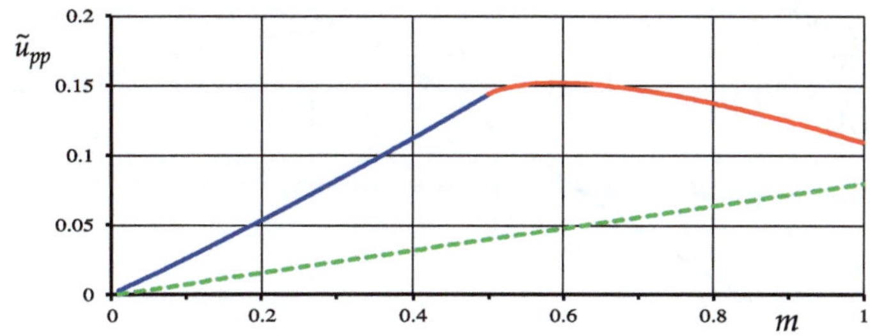

Fig. 6.2 Normalized peak-to-peak PV voltage as a function of modulation index in case of sinusoidal current with unity power factor (I_{ac} = 1 A). Basic H-bridge configuration (dashed line) and H-bridge with LDN (continuous lines)

trace) and for the H-bridge with LDN (bleu and red traces). In case of H-bridge with LDN configuration, it can be noted that the normalized peak-to-peak PV voltage ranges between 0 and 0.15 ≅ 3/20. The absolute maximum of PV voltage becomes:

$$\tilde{v}_{pp}\Big|^{\max} = \frac{3}{20}\frac{I_{ac}}{fC_H} \tag{6.12}$$

If the maximum low-frequency PV voltage oscillations is assigned, the dc-link capacitor can be sized by reversing (6.12):

$$C_H \geq \frac{3}{20}\frac{I_{ac}}{f\tilde{v}_{pp}\Big|^{\max}} \tag{6.13}$$

Comparing the normalized peak-to-peak PV voltage oscillations in case of basic (H-bridge) and multilevel (H-bridge with LDN) inverter configurations, the dc-link capacitor in case of H-bridge with LDN should be doubled, according to Fig. 6.2 in order to have the same peak-to-peak PV voltage oscillations.

6.2.3 PV Current Analysis

As for the PV voltage v, the instantaneous PV current i can be written in terms of components, corresponding to (6.4). Neglecting the switching frequency component, it becomes:

$$i = I + \tilde{i} \tag{6.14}$$

being I the average over the fundamental period (dc component) and $\tilde{i} = \tilde{v}/R_{PV}$ the alternating low-frequency harmonic component. Note that, due to the steady-state behavior of dc-link capacitor, $I = I^H$.

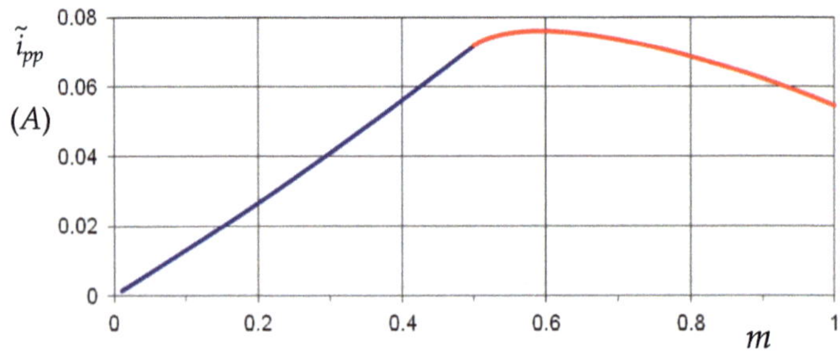

Fig. 6.3 Peak-to-peak PV current as a function of modulation index in case of I_{ac} = 1 A (sinusoidal current with unity power factor), R_{PV} = 40 Ω, and C_H = 1 mF

Basing on the peak-to-peak PV voltage given by (6.10) and (6.11), the peak-to-peak PV current ($\tilde{i}_{pp} = \tilde{v}_{pp}/R_{PV}$) can be written as:

$$\tilde{i}_{pp} = \begin{cases} \frac{I_{ac}}{R_{PV}\omega C_H}\left[-m\arcsin\left(\frac{m}{2}\right)+\frac{m\pi}{2}+\frac{m^2}{2}\sqrt{4-m^2}\right], & m \leq 0.5 \\ \frac{I_{ac}}{R_{PV}\omega C_H}\left[2m\arcsin\left(\frac{1}{2m}\right)-\frac{m\pi}{2}+\sqrt{1-\frac{1}{4m^2}}-m\arcsin\left(\frac{m}{2}\right)+\frac{m^2}{2}\sqrt{4-m^2}\right], & m \geq 0.5 \end{cases} \tag{6.15}$$

As an example, Fig. 6.3 shows the peak-to-peak PV current low-frequency component as a function of modulation index in case of dc-link capacitor C_H = 1 mF and unity sinusoidal output current with unity power factor, considering a PV series equivalent resistance R_{PV} = 40 Ω.

According to (6.10), (6.11) and (6.15), it is clear that the amplitudes of the PV voltage and current oscillations are inversely proportional to the dc-link capacitor C_H since the dc-link capacitor reactance is dominating the dc-link impedance. However, amplitude of PV current oscillations also depends on R_{PV}, which is a function of the operating point on the PV field characteristic. Considering a PV plant of few kW (single-phase), R_{PV} is ranging between few Ω near the open-circuit point, passing around few tens Ω near the MPP, up to hundreds or even thousands Ω toward the short-circuit point [7]. These considerations can be readily exploited to design the dc-link capacitor C_H in order to obtain PV voltage and current oscillation amplitudes in the order of few % comparing to the rated MPP values.

6.3 RCC-MPPT Algorithms in Case of Single and Multiple PV Harmonics

The block diagram of the whole PV control scheme is depicted in Fig. 6.4. The RCC algorithm is used to estimate the voltage derivative of the power, *dP*/*dV*, driving the working point toward the MPP and determining the reference grid

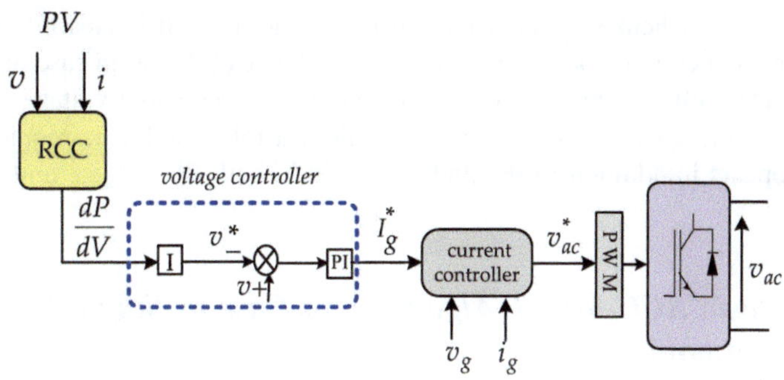

Fig. 6.4 Block diagram of the whole PV control scheme

current amplitude I_g^*. In particular, the reference dc-link voltage v^* is simply obtained by integrating the estimated dP/dV, I_g^* is determined by a PI voltage regulator, and dq current controller has been implemented in order to inject a sinusoidal grid current with unity power factor.

In order to provide for a comparative example of PV voltage and current oscillations in case of basic (H-bridge) and multilevel (H-bridge with LDN) inverter configurations, Fig. 6.5 shows the results of a numerical simulation considering the

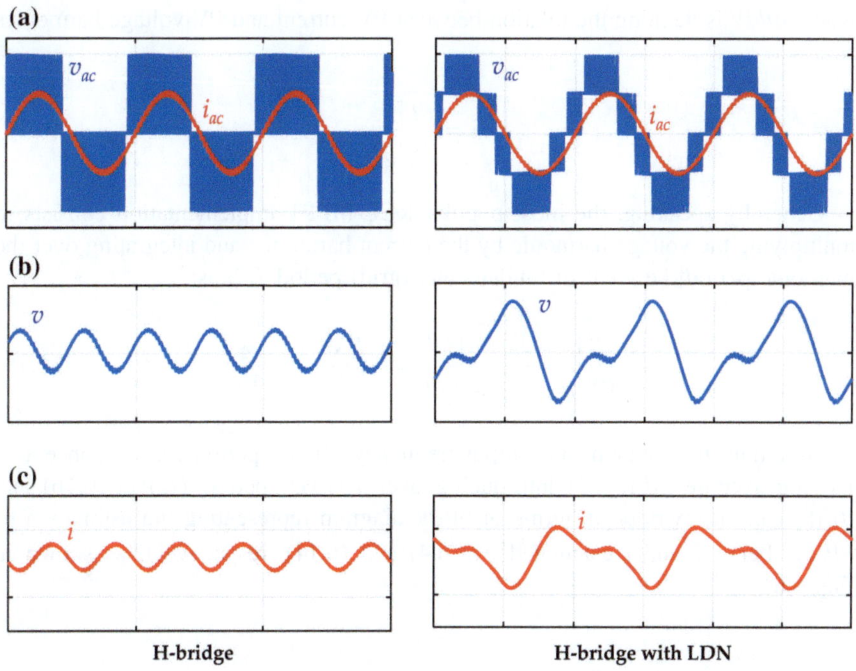

Fig. 6.5 Simulation examples of: **a** grid current and inverter output voltage, **b** PV voltage, and **c** PV current, in case of H-bridge without LDN (left column), and in case of H-bridge with LDN (right column)

two conversion schemes in the same working conditions. It is clearly visible that PV voltage and current contain only the 2nd harmonic (100 Hz) in case of H-bridge without LDN (left column). Increasing the number of the output voltage level from 3 to 5 with the considered LDN configuration, a relevant 1st harmonic (50 Hz) clearly appears in addition to the 2nd harmonic in both PV voltage and current.

6.3.1 Basic RCC-MPPT Algorithm in Case of Single PV Harmonic

In case of single-phase single-stage PV systems, being the inverter directly connected to the PV field, inherent 2nd order harmonic of instantaneous power appears in PV current and PV voltage. As known, ripple correlation control algorithm uses these oscillations for providing information about the operating point of the PV field. For the considered working point Q, the voltage derivative of the PV power dP/dV is written as:

$$\left.\frac{dP}{dV}\right|_Q = I + \left.\frac{dI}{dV}\right|_Q V \tag{6.16}$$

where dI/dV is defining the relation between PV current and PV voltage harmonics as:

$$\tilde{i} = \left.\frac{dI}{dV}\right|_Q \tilde{v} \tag{6.17}$$

Generally speaking, the most popular RCC-MPPT implementation consists in multiplying the voltage harmonic by the current harmonic, and integrating over the harmonic period, i.e., half of fundamental (grid) period $T/2$, as:

$$\left.\frac{dI}{dV}\right|_Q = \frac{\int_{t-T/2}^{t} \tilde{i}\tilde{v}dt}{\int_{t-T/2}^{t} \tilde{v}^2 dt} = -\frac{I_2 V_2}{V_2^2} = -\frac{I_2}{V_2} \tag{6.18}$$

Assuming the fundamental output frequency 50 Hz (period T), the concept of moving average ($MAvg$) is introducing over 100 Hz (period $T/2$) in (6.16) and (6.18) can be rewritten in terms of block diagram representing the estimation of dP/dV, being $\tilde{v}$ and $\tilde{i}$ calculated by (6.4) and (6.14), respectively, as shown in Fig. 6.6.

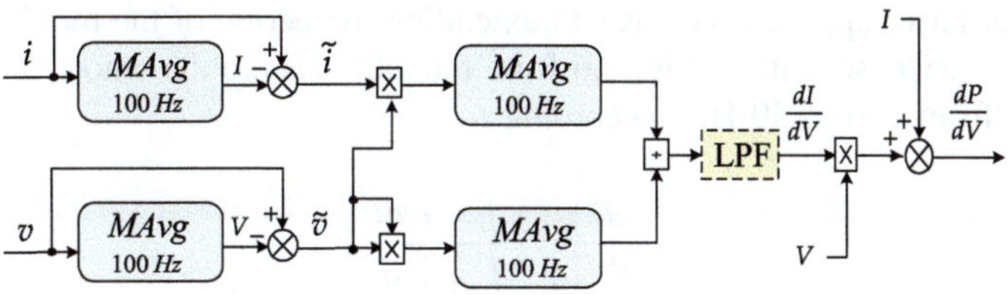

Fig. 6.6 Block diagram of the basic RCC algorithm to estimate *dP*/*dV* in case of a single harmonic component using *MAvg* over 100 Hz (2nd harmonic)

6.3.2 Proposed RCC-MPPT Algorithm in Case of Multiple PV Harmonics

In case of multiple PV voltage and current harmonics (multilevel topology), the estimation of *dI*/*dV* (and then *dP*/*dV*) should be carried out by considering a specific harmonic (*k*th order), preferably the one with highest amplitude to increase the resolution. In this case, (6.17) can be rewritten as:

$$\tilde{i}_k = \left.\frac{dI}{dV}\right|_Q \tilde{v}_k \tag{6.19}$$

By following the same approach as in previous section, (6.18) can be summarized for each considered harmonic as:

$$\left.\frac{dI}{dV}\right|_Q^k = \frac{\int_{t-T}^{t} \tilde{i}_k \tilde{v}_k dt}{\int_{t-T}^{t} \tilde{v}_k^2 dt} = -\frac{I_k V_k}{V_k^2} = -\frac{I_k}{V_k} \tag{6.20}$$

Note that integrals in Eq. (6.20) are now calculated over the fundamental period *T* (corresponding to the grid frequency), as defined by Fourier series to calculate the *k*th harmonic. The implementation of Eqs. (6.16) and (6.20) is shown in the block diagram of Fig. 6.7, estimating the *dP*/*dV* considering only the *k*th harmonic component.

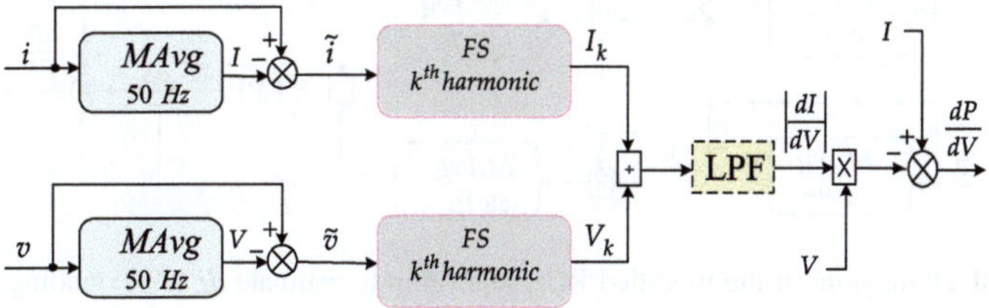

Fig. 6.7 Block diagram of proposed RCC estimating *dP*/*dV* by Fourier series (FS) with a single harmonic

An alternative approach consists in extending the period of the moving average in the basic RCC scheme of Fig. 6.6 from the half (*T*/2) to the whole fundamental period *T* (from 100 to 50 Hz), according to:

$$\left.\frac{dI}{dV}\right|_Q = \frac{\int_{t-T}^{t} \tilde{i}\tilde{v}dt}{\int_{t-T}^{t} \tilde{v}^2 dt} \tag{6.21}$$

The corresponding block diagram to estimate *dP*/*dV* is presented in Fig. 6.8.

The relationship between the single harmonic method given by (6.20) and the extended moving average method (6.21) can be carried out by considering all the harmonics of the alternating PV voltage and current components:

$$\tilde{v} = \sum_k \tilde{v}_k, \quad \tilde{i} = \sum_k \tilde{i}_k \tag{6.22}$$

In fact, considering the Parseval's theorem and introducing (6.22) in (6.21), the estimation of *dI*/*dV* can be written as:

$$\left.\frac{dI}{dV}\right|_Q = \frac{\int_{t-T}^{t} \sum_k \tilde{i}_k \sum_k \tilde{v}_k dt}{\int_{t-T}^{t} \left(\sum_k \tilde{v}_k\right)^2 dt} = -\frac{\sum_k I_k V_k}{\sum_k V_k^2} \tag{6.23}$$

Equation (6.23) can be reorganized multiplying and dividing by V_k, as follows:

$$\left.\frac{dI}{dV}\right|_Q = -\frac{\sum_k \left(V_k^2 \frac{I_k}{V_k}\right)}{\sum_k V_k^2} = \frac{\sum_k \left(w_k \left.\frac{dI}{dV}\right|_Q^k\right)}{\sum_k w_k} \tag{6.24}$$

Equation (6.24) states that the estimation of *dI*/*dV* made by the modified moving average method (6.21) (i.e., *MAvg* over the fundamental period *T*) is equivalent to the weighted average of the estimation of *dI*/*dV* obtained by the individual harmonics (6.20), using as weight (w_k) the square of the *k*th voltage harmonic amplitude:

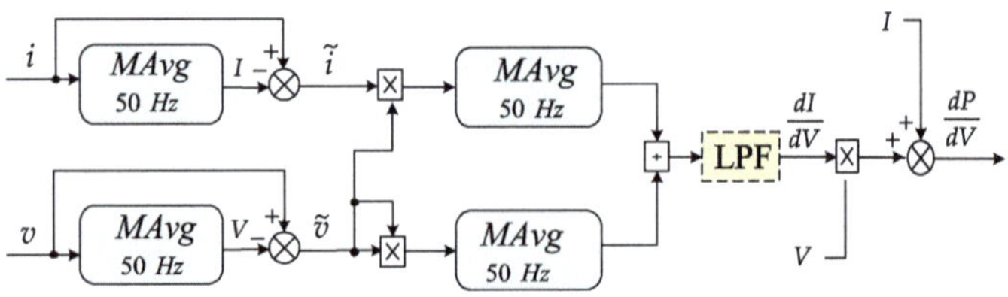

Fig. 6.8 Block diagram of the modified RCC algorithm to estimate *dP*/*dV* extending *MAvg* over *T* (50 Hz)

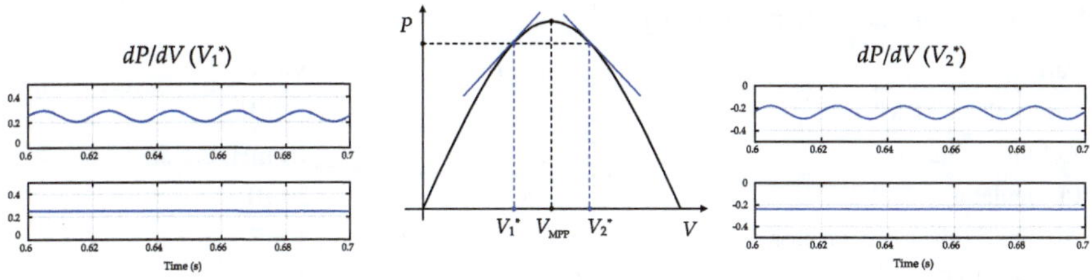

Fig. 6.9 Simulation results: steady-state estimation of *dP*/*dV* by RCC with moving average over different periods and working points on the left ($V_1^* = 45$ V) and on the right ($V_2^* = 55$ V) of the MPP ($V_{MPP} = 50$ V). Top traces: *MAvg* over *T*/2 (100 Hz, basic RCC), and bottom traces: *MAvg* over *T* (50 Hz, modified RCC)

$$w_k = V_k^2 \tag{6.25}$$

In order to preliminarily prove the effectiveness of modified RCC comparing to basic RCC for the estimation of *dP*/*dV*, a simulation example is given in Fig. 6.9.

In particular, considering a simple linear *I*-*V* curve corresponding to a parabolic *P*-*V* curve, the *dP*/*dV* is calculated by the basic RCC scheme of Fig. 6.6 (*MAvg* over *T*/2, 100 Hz, top), and the modified RCC scheme of Fig. 6.8 (*MAvg* over *T*, 50 Hz, bottom), in case of H-bridge with LDN (i.e., Figs. 6.1, and 6.5, right column). Reference is made to the two steady-state operating points symmetrically placed on the left- and on the right-side of the MPP, corresponding to $dP/dV \cong \pm 0.25$ A, as depicted in Fig. 6.9. It is evident that applying the *MAvg* at 100 Hz results in an oscillating estimation of *dP*/*dV* whereas the estimation is precise and stable applying the *MAvg* at 50 Hz, as a consequence of Eq. (6.21).

6.4 Simulations and Comparative Analysis

Simulation results have been carried out in order to prove the theoretical developments presented in previous sections. Reference is made to the multilevel inverter configuration obtained by cascading H-bridge and LDN, as in Fig. 6.1. Single phase grid-connected photovoltaic systems have been numerically implemented by MATLAB-Simulink on the basis of the parameters shown in Table 6.1. A string of series/parallel connected PV modules has been adopted as PV source, according to the data given in Table 6.2 (STC). In particular, PV modules are SP-305 type (96 cells, monocrystalline).

Figure 6.10 shows the simulation results obtained by estimating *dP*/*dV* by (6.16) and (6.21), i.e. *MAvg* over 50 Hz, as in Fig. 6.8, when an irradiance transient occurs (Fig. 6.10a). It can be noted that the estimation is correct and stable in steady-state. However, as generally known for RCC, it is oscillating during the irradiance

Table 6.1 Simulated system parameters

Parameter	Symbol	Value
Grid voltage (RMS)	V_g	230 V
AC-link inductor	L_f, R_f	10 mH, 0.1 Ω
PI regulator	K_P, K_I	20,10
AC-link H-bridge capacitor	C_H	5 mF
DC-link LDN capacitor	C_L	20 mF
Switching frequency	f_{sw}	2 kHz

Table 6.2 PV array specifications (STC)

Parameter	Symbol	Value
Open circuit voltage	V_{OC}	64.2 V
Short circuit current	I_{SC}	5.96 A
Maximum photovoltaic voltage	V_{MPP}	54.7 V
Maximum photovoltaic current	I_{MPP}	5.58 A
Number of series-connected modules per string	N_S	9
Number of parallel strings	N_P	3

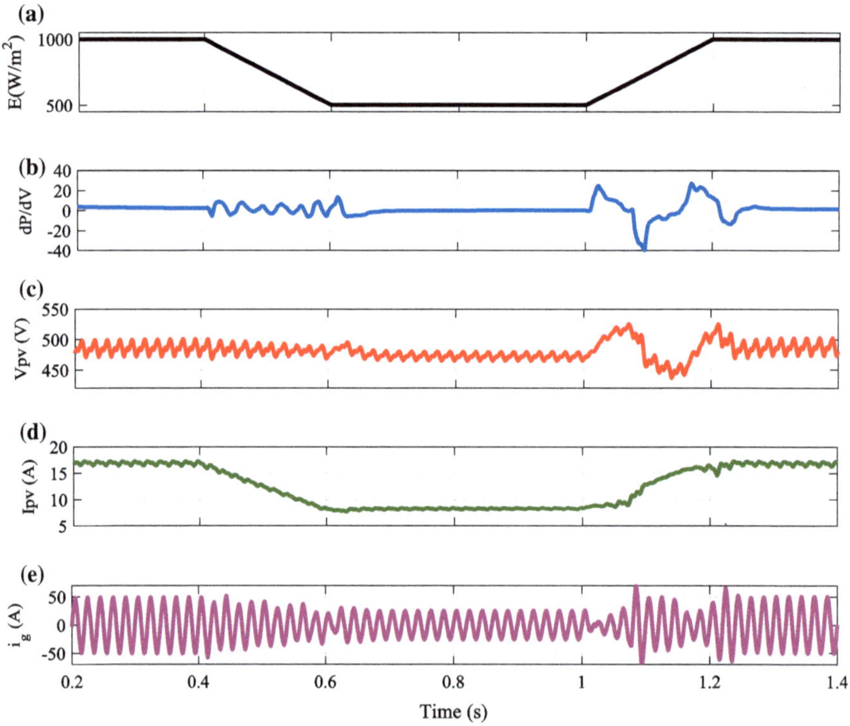

Fig. 6.10 Irradiance transient in case of RCC-MPPT algorithm using *MAvg* over 50 Hz (Fig. 6.8): **a** irradiance, **b** estimated *dP/dV*, **c** PV voltage, **d** PV current, and **e** grid current

transients, introducing dc-link voltage overshoots (Fig. 6.10c), acceptable PV current transients, but large grid current fluctuations (Fig. 6.10e).

The *P*(*V*) trajectory of the previous transient is presented in Fig. 6.11, together with the static characteristics of the PV field. The large transient corresponding to the increasing irradiance ramp is well visible.

Figure 6.12 shows the simulation results obtained by estimating dP/dV by (6.16) and (6.20), i.e. considering only a specific harmonic order, as in Fig. 6.7, in this case the first harmonic (50 Hz) is considered. It can be noted that the estimation is correct and stable both in steady-state and during the irradiance transients, with very small and acceptable perturbations (Fig. 6.12b) just at the beginning and at the end of the irradiance ramps (Fig. 6.12a). The resulting PV voltage and current are smoothed (Fig. 6.12c, d), and the grid current changes its amplitude between the steady-state values with a profile corresponding to the irradiance ramp, practically without any further transient (Fig. 6.12e).

The *P*(*V*) trajectory of the previous transient is presented in Fig. 6.13, together with the static characteristics of the PV field. In this case, the working point straightly moves between the steady-state MPPs, without any oscillations.

Similar results can be obtained by the Fourier Series method also selecting different harmonics order. In any case, harmonics amplitude should be significant in order to maintain a good resolution in the estimation of dP/dV. In the considered case of H-bridge with LDN inverter, the other noticeable harmonic is only at 100 Hz, as clearly visible in Fig. 6.5.

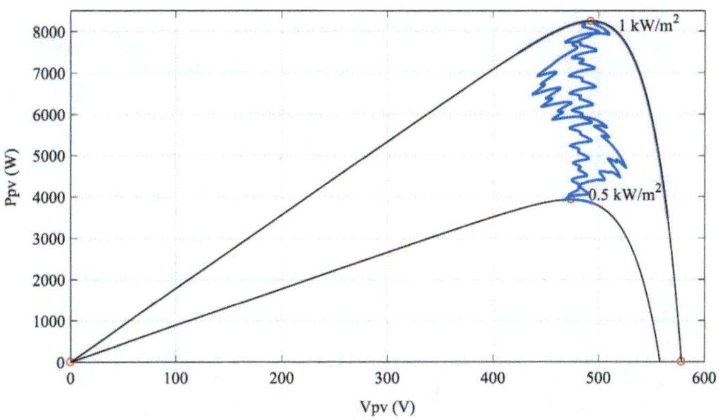

Fig. 6.11 Trajectory of the working point on *P*(*V*) plane in case of RCC-MPPT algorithm using *MAvg* over 50 Hz, corresponding to Fig. 6.10

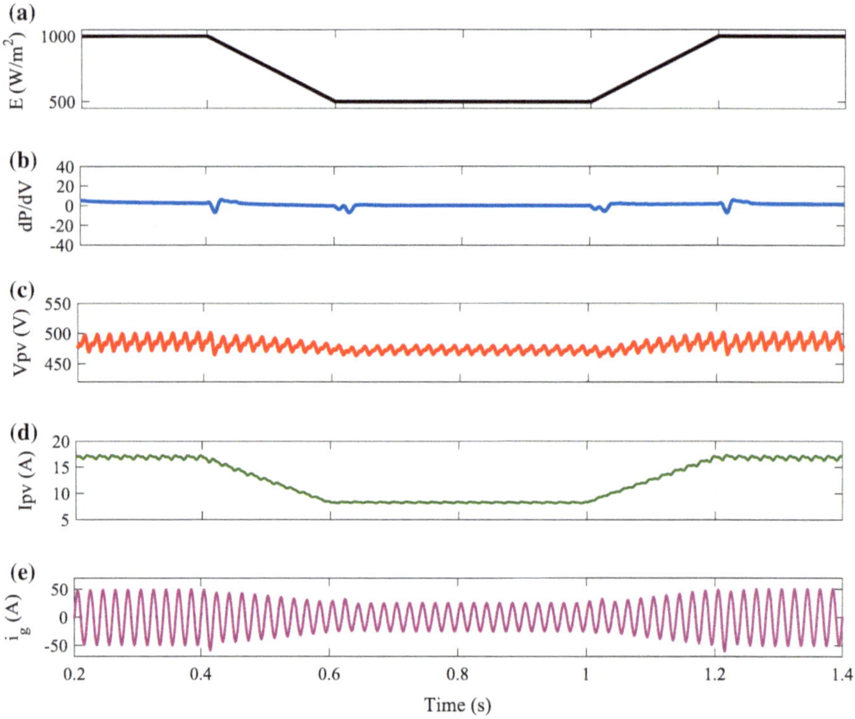

Fig. 6.12 Irradiance transient in case of RCC-MPPT algorithm using only the 50 Hz harmonic components (FS, Fig. 6.7): **a** irradiance, **b** estimated dP/dV, **c** PV voltage, **d** PV current, and **e** grid current

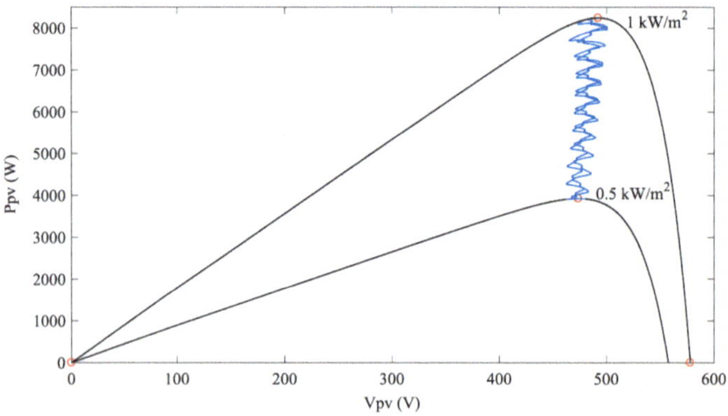

Fig. 6.13 Trajectory of the working point on $P(V)$ plane in case of RCC-MPPT algorithm using only the 50 Hz components (FS), corresponding to Fig. 6.12

6.5 Implementation and Experimental Results

In order to verify the effectiveness of the multilevel PV inverter with the proposed RCC-MPPT algorithms in both steady-state and transient conditions, a grid-connected PV generation system with H-bridge and LDN has been implemented, according to the circuit scheme of Fig. 6.14.

The picture of the corresponding experimental setup is shown in Fig. 6.15. It consists of two power boards (H-bridge and LDN) based on Mitsubishi IGBT smart modules, IPM PS22A76 (1200 V, 25 A), Yokogawa DLM 2024 oscilloscope with the PICO TA057 differential voltage probe (25 MHz, ±1400 V, ±2%) and LEM PR30 current probe (dc to 20 kHz, ±20 A, ±1%). Additionally, two current transducers model LA 55-P (LEM® company) were used to measure PV and grid currents, while PV and grid voltages were measured using two voltage transducers model LV 25-P (LEM® company). The switching frequency of the multilevel grid-connected inverter is set to 2.5 kHz. For digital implementation of current and voltage controllers, as well as the PLL and the RCC-MPPT algorithms, a Digital signal processor DSP (DSP TMS320F28377D) was used to generate the PWMs signals for the H-bridge and the LDN boards of the multilevel LDN inverter. An optical interface board links DSP with power boards. Results are shown by oscilloscope screenshots, elaborating and emphasizing the signals of interest. The main parameters of the experimental setup are given in Table 6.3.

The DC source has been implemented by a resistive voltage supply with variable series resistance for preliminary steady-state and transient tests, according to the scheme of Fig. 6.14. The corresponding parameters are given in Table 6.4.

For realistic tests, a reduced-scale array consisting of two series-connected PV modules has been adopted, introducing different irradiance conditions by covering/uncovering with a white leave, well representing sunny and cloudy conditions (the

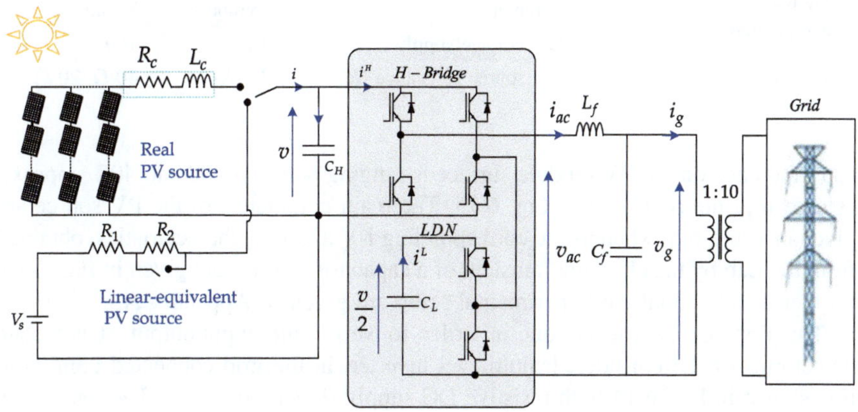

Fig. 6.14 Circuit scheme of the experimental setup including the whole multilevel PV generation system

Fig. 6.15 Picture view of the experimental setup in the Lab (left side) and arrangement of the PV array (two modules) on the roof, free and shadowed by a white sheet (≅500 and 200 W/m^2, right side)

Table 6.3 System parameters

Parameter	Symbol	Value
Dc-link H-bridge capacitance	C_H	2.2 mF
Dc-link LDN capacitance	C_L	3.3 mF
AC filter capacitance	C_f	25 μF
AC filter resistance and inductance	R_f, L_f	1.1 Ω, 8.8 mH
Grid voltage (RMS) (1:10 transformer)	V_g	22/220 V
Grid (fundamental) frequency	f	50 Hz

Table 6.4 Resistive DC source parameters

Parameter	Symbol	Value
DC voltage supply	V_S	100 V
DC source resistance (R_S)	R_1, R_2	40 Ω, 20 Ω

sun irradiance on the PV module surface is ranging between 100 and 40% approx), as shown in the right side of Fig. 6.15. The main parameters of the PV source are given in Table 6.5, whereas the corresponding I-V and P-V characteristics, obtained from the Lab by the charging transient of a capacitor (1 mF), are given in Fig. 6.16. Further details about the experimental setup are given in Appendix A.

The first test is carried out in order to verify the input/output steady-state waveforms of the considered multilevel inverter, in the grid-connected configuration shown in Fig. 6.14 with resistive DC supply (linear-equivalent PV source). In particular, Fig. 6.17a shows grid voltage and current (top half screen) and PV voltage and current (bottom half screen), whereas Fig. 6.17b shows the time zoom

Table 6.5 PV source parameters—Standard test conditions (STC)

Parameter	Symbol	Value
Open circuit voltage	V_{OC}	43.4 V
Short circuit current	I_{SC}	4.8 A
Maximum photovoltaic voltage	V_{MPP}	34 V
Maximum photovoltaic current	I_{MPP}	4.4 A
Number of series-connected PV modules	N_S	2
Resistance and inductance of PV cable	R_c, L_c	0.5 Ω, 40 μH

Fig. 6.16 *P-V* and *I-V* characteristics of the PV array including the cable with an irradiance of 500 W/m^2 (top) and 200 W/m^2 (white sheet, bottom) and $T_{cell} \cong 35$ °C (reference is made to Fig. 6.15)

of inverter voltage and current (top half screen) and PV voltage and current (bottom half screen). As expected, the inverter voltage has a proper multilevel waveform over 5 levels, and inverter (grid) current is almost sinusoidal despite the switching frequency is only 2.5 kHz, with unity power factor. PV voltage and PV current have oscillations including both 1st and 2nd harmonic components (50 and 100 Hz, respectively), in phase opposition.

The following tests are carried out in order to verify the dynamic performance of the different RCC-MPPT algorithms in case of irradiance transients. In particular, two sets of experimental tests have been considered. First, the linear-equivalent PV source has been selected in order to simulate extremely fast irradiance transients by switching on/off the series resistances, performing dynamic comparative tests for all the four types of RCC-MPPT algorithms. Last tests are performed considering the real PV source, introducing irradiance transients by shadowing/unshadowing the PV modules by a white sheet, well representing the effects of real clouds (reference to Figs. 6.15 and 6.16).

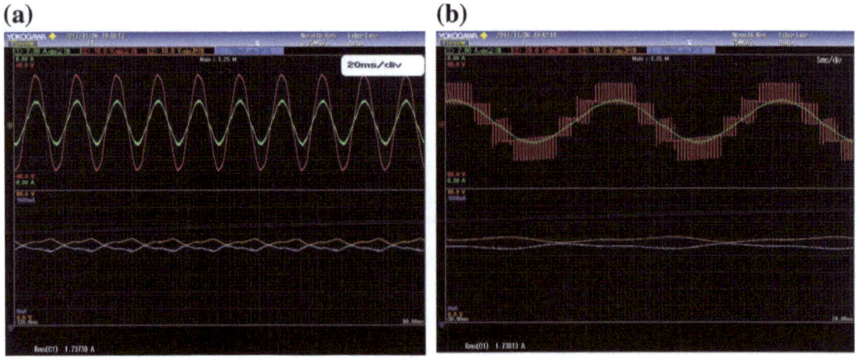

Fig. 6.17 Input/output steady-state waveforms of the converter with $V_S = 100$ V, $R_S = 60\ \Omega$, $V = V_{MPP} = 50$ V: **a** grid voltage and current (top traces), PV voltage and current (bottom traces), **b** time zoom: inverter voltage and current (top traces), PV voltage and current (bottom traces)

Figure 6.18 shows the grid voltage and current as well as the PV current and voltage in case of transients obtained by switching on/off the series resistances (Fig. 6.14; Table 6.4), representing a step irradiance transient. Diagrams in Fig. 6.18a are based on the basic RCC scheme with *MAvg* over 100 Hz (corresponding to Fig. 6.6), while diagrams in Fig. 6.18b are based on the modified RCC scheme with *MAvg* over 50 Hz (corresponding to Fig. 6.8).

In both cases, as expected, the steady-state MPP voltage is equal to the half of the DC source voltage ($V_S = 100$ V, $V_{MPP} = 50$ V). In this case, the (equivalent) irradiance step is seen as a perturbation by the PV voltage controller. Correspondingly, the PV current shows a fast step transient. Despite the basic RCC scheme gives acceptable results, it leads to larger overshoots and higher settling times comparing to the modified RCC.

Figure 6.19 shows the same quantities with the same kind of transients as in Fig. 6.18, but referred to the modified RCC schemes employing only a specific harmonic component (reference is made to Fig. 6.7).

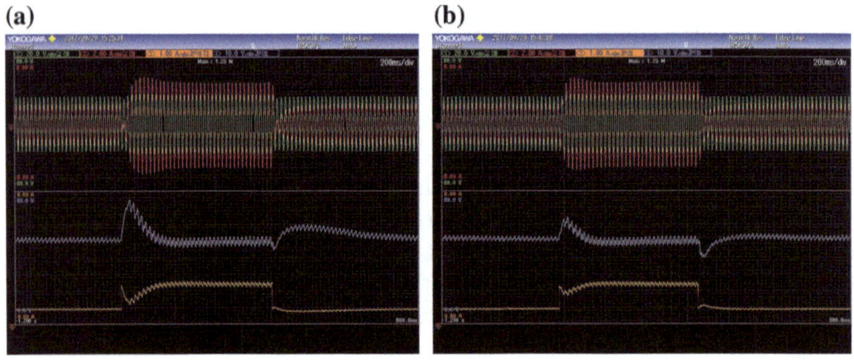

Fig. 6.18 Experimental results of RCC-MPPT algorithms in case of step transients of R_S. Top traces: grid current and voltage. Bottom traces: PV current and voltage. **a** *MAvg* over 100 Hz and **b** *MAvg* over 50 Hz

In particular, diagrams in Fig. 6.19a are obtained by the modified RCC scheme based on the 2nd harmonic component (100 Hz), whereas diagrams in Fig. 6.19b are obtained by the modified RCC scheme based on the 1st harmonic component (50 Hz). Both these modified RCC-MPPT schemes give similar results, being similar the amplitude of voltage and current harmonics with smoothed overshoots and short settling times for the PV variables. In particular, the modified RCC scheme employing the 1st harmonic component (50 Hz) seems the more effective, also offering the advantage of using the highest harmonic amplitude in the typical modulation index range for grid-connected PV generation schemes (i.e. *m* between 0.7 and 0.8).

For these reasons, only this last modified RCC-MPPT scheme is considered in the last tests presented in Fig. 6.20, considering a real PV source (2 PV modules),

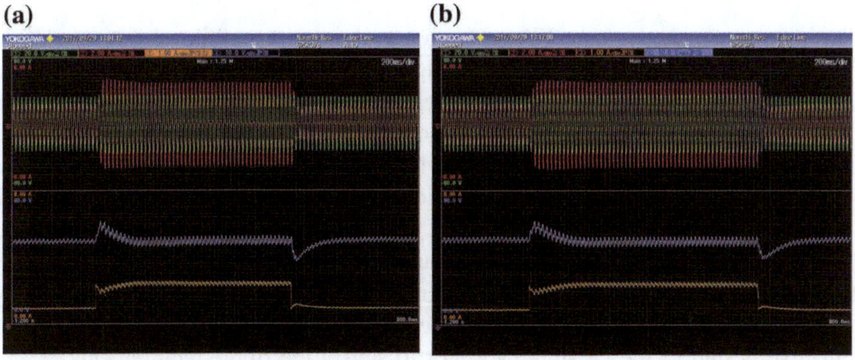

Fig. 6.19 Experimental results of modified RCC-MPPT algorithms in case of step of R_S. Top traces: grid voltage and current. Bottom traces: PV voltage and current. **a** Using only 2nd harmonic (100 Hz), and **b** using only 1st harmonic (50 Hz)

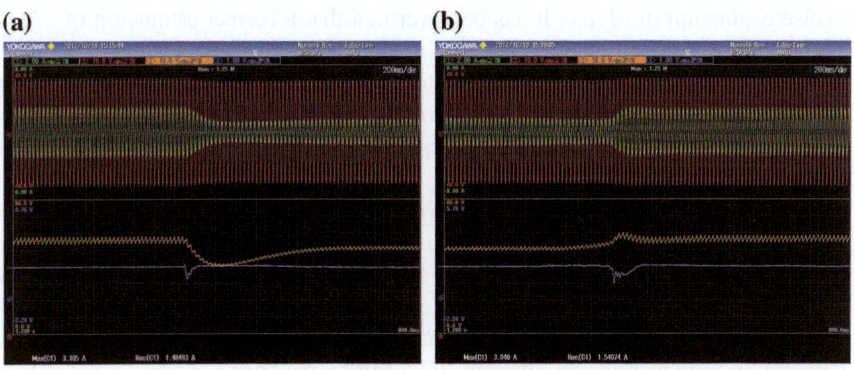

Fig. 6.20 Decrease and increase sun irradiance transients (500–200 W/m^2) using only the 50 Hz harmonic component (Fourier)

in real environmental operating conditions, and with realistic irradiance transients. Increasing and decreasing sun irradiance transients obtained by shadowing and unshadowing the PV modules by a white sheet have been considered, corresponding to the P-V and I-V characteristics of Fig. 6.16 (500 and 200 W/m^2). In particular, Fig. 6.20 shows grid voltage and current (top half screen), together with PV voltage and the estimation of dP/dV (bottom half screen). As expected, the estimation of dP/dV fails during the initial part of the transient, but without to introduce particular drawbacks in PV and grid variables. In fact, both for increasing and decreasing irradiance transients, grid current amplitude has a smoothed and fast profile, without significant overshoots.

6.6 Discussion

A single-phase single-stage PV generation system with multilevel output voltage waveforms and improved RCC-MPPT algorithm has been proposed and examined in details. The multilevel inverter is implemented by an LDN cell, as kind of retrofit to the basic H-bridge cell, increasing the output voltage levels from three to five. In addition to the 2nd harmonic resulting from the basic H-bridge configuration, a multilevel configuration introduces additional voltage and current harmonics on the input (PV) side. In particular, a relevant 1st order harmonic is noticeable, the 3rd harmonic is slightly appreciable, whereas higher order harmonics are completely negligible. Both PV voltage and current harmonic amplitudes are analytically calculated in the whole modulation range, offering the possibility of a precise and effective design of the DC-link capacitor to satisfy the ripple requirements.

Due to the additional harmonics, the basic RCC-MPPT scheme becomes inadequate, leading to a misestimation of dP/dV. A modified RCC scheme extracting the amplitude of a specific harmonic form PV voltage and current waveforms has been proposed in order to overcome this drawback. Reference has been made to the harmonic with highest amplitude in order to maximize the resolution, leading to a precise estimation of dP/dV. It has been verified that a correct estimation of dP/dV can be simply obtained by doubling the time window of low-pass filters in the RCC scheme (i.e., moving average, from $T/2$ to T). It has been proved that the resulting dP/dV is a weighted average of all the dP/dV estimations performed by the individual PV voltage and current harmonics, being the weight the individual voltage harmonic amplitude itself.

Numerical and experimental tests have been carried out to prove the effectiveness of the whole PV generation scheme, including multilevel waveforms and improved RCC-MPPT algorithms. Both linear-equivalent PV source and real PV source have been implemented in the experimental setup, considering both steady-state and fast sun irradiance transient conditions in order to verify the dynamic performance of the different RCC-MPPT methods.

References

1. C. Barth, R.C.N. Pilawa-Podgurski, Dithering digital ripple correlation control for photovoltaic maximum power point tracking. IEEE Trans. Power Electron. **30**(8), 4548–4559 (2015)
2. A.M. Bazzi, P.T. Krein, Ripple correlation control: an extremum seeking control perspective for real-time optimization. IEEE Trans. Power Electron. **29**(2), 988–995 (2014)
3. D. Casadei, G. Grandi, C. Rossi, Single-phase single-stage photovoltaic generation system based on a ripple correlation control maximum power point tracking. IEEE Trans. Energy Convers. **21**(2), 562–568 (2006)
4. C. Boonmee, Y. Kumsuwan, Control of single-phase cascaded H-bridge multilevel inverter with modified MPPT for grid-connected photovoltaic systems, in *IECON 2013, 39th Annual Conference of the IEEE Industrial Electronics Society* (2013), pp. 566–571
5. M. Hammami, G. Grandi, M. Rudan, An improved MPPT algorithm based on hybrid RCC scheme for single-phase PV systems, in *IECON 2016—42nd Annual Conference of the IEEE Industrial Electronics Society* (2016), pp. 3024–3029
6. C. Boonmee, Y. Kumsuwan, Modified maximum power point tracking based-on ripple correlation control application for single-phase VSI grid-connected PV systems, in *2013 10th International Conference on Electrical Engineering/Electronics, Computer, Telecommunications and Information Technology, ECTI-CON 2013* (2013), pp. 1–6
7. T. Suntio, T. Messo, A. Aapro, J. Kivimäki, A. Kuperman, Review of PV generator as an input source for power electronic converters. Energies **10**(8), 1076 (2017)

Authored Papers

8. M. Hammami, G. Grandi, M. Rudan, An improved MPPT algorithm based on hybrid RCC scheme for single-phase PV systems, in *42nd Annual Conference of the IEEE Industrial Electronics Society, IECON 2016*, Florence, Italy, 24–27 Oct 2016
9. M. Hammami, G. Grandi, Single-phase multilevel PV generation system with improved ripple correlation control MPPT algorithm. Energies **10**(12), 2037 (2017)
10. M. Hammami, M. Vujacic, A. Viatkin, G. Grandi, Analysis of a flexible single-phase multilevel inverter topology for photovoltaic applications, in *9th International Renewable Energy Congress, IREC 2018*, Hammamet, Tunisia, 20–22 March 2018

Chapter 7
Conclusion and Perspectives

7.1 Conclusion

This thesis deals with the improvement of the stability and the dynamic performance of Ripple Correlation Control (RCC) Maximum Power Point Tracking (MPPT) algorithms in case of grid-connected inverters with particular reference to the transient behavior. In order to improve the power quality, a new multilevel inverter topology with level doubling network (LDN) is introduced. It is based on a modular half-bridge (two switches) cascaded to a full H-bridge in order to double the output voltage levels. The improvement of the RCC scheme has been extended to the case of multilevel inverters since additional PV voltage and current low-frequency harmonics are introduced, perturbing the basic implementation of the RCC scheme, leading to malfunctioning. In order to estimate the voltage derivative of power (dP/dV) and to drive the PV operating point to the Maximum Power Point (MPP) in a more precise and faster manner, analytical calculation of voltage and current harmonic components used in RCC schemes has been presented and different possible dP/dV estimations have been proposed in case of multiple harmonics, and a specific case with LDN has been considered. The expressions of the peak-to-peak dc-link current and voltage ripple amplitudes have been analytically determined as a function of the modulation and effective normalization has been introduced to better design the dc-link capacitors of both H-bridge and LDN cells basing on the desired dc-link voltage ripple requirements. The analysis is accompanied with the numerical simulation and experimental results. Experimental verifications of the LDN inverter and also of RCC-MPPT algorithms in case of multiple harmonics have been carried out.

M. Hammami, *Level Doubling Network and Ripple Correlation Control MPPT Algorithm for Grid-Connected Photovoltaic Systems*, Springer Theses,
https://doi.org/10.1007/978-3-030-10492-4_7

7.2 Perspectives

A significant new knowledge has been developed and presented in this Ph.D. thesis. However, a large space is left for further research works. Some of them are listed here, as possible realistic perspectives:

- the extension to the three-phase multilevel conversion systems can be further developed, including the performance of the RCC-MPPT;
- hardware implementation of a three-phase conversion system, including a RCC-MPPT algorithm, to investigate the possibility of extension to medium-high power PV plants;
- investigation of the possibility to apply the multiple harmonics RCC to other multilevel inverter configurations (both single-and three-phase);
- extension of the analysis to any phase angle between grid voltage and inverter current (any output power factor);
- investigation of different multilevel modulation strategies in case of three-phase inverters.

Appendix A
Experimental Setup

A.1 General Description

All experimental activities related to the design and practical realization of the considered configurations were carried out in the laboratory "SolarTronic ST-Lab" at the department of Electrical, Electronic and Information Engineering "Guglielmo Marconi", University of Bologna and headed by Professor Gabriele Grandi. The picture of the experimental setup of a grid-connected PV generation system with H-bridge and LDN is shown in Fig. A.1. It consists of two power boards (H-bridge and LDN) based on Mitsubishi IGBT smart modules, an IPM PS22A76 (1200 V, 25 A, Mitsubishi Electric Corporation, Tokyo, Japan), a Yokogawa DLM 2024 oscilloscope (Yokogawa Electric Corporation, Tokyo, Japan) with the PICO TA057 differential voltage probe (25 MHz, ±1400 V, ±2%, Pico Technology, Tyler TX, USA) and LEM PR30 current probe (dc to 20 kHz, ±20 A, ±1%, LEM Europe GmbH, Fribourg, Switzerland). Additionally, two current transducers model LA 55-P (LEM® company) were used to measure PV and grid currents, while the PV and grid voltages were measured using two voltage transducers model LV 25-P (LEM® company). For the digital implementation of current and voltage controllers, as well as the PLL and the RCC-MPPT algorithms, a Digital signal processor DSP (DSP TMS320F28377D) was used to generate the PWMs signals for the H-bridge and the LDN boards of the multilevel LDN inverter. An optical interface board links DSP with power boards.

A.2 PV Module

For realistic tests, a reduced-scale array consisting of two series-connected PV modules has been adopted, introducing different irradiance conditions by covering/uncovering with a white leave, well representing sunny and cloudy

M. Hammami, *Level Doubling Network and Ripple Correlation Control MPPT Algorithm for Grid-Connected Photovoltaic Systems*, Springer Theses,
https://doi.org/10.1007/978-3-030-10492-4

Fig. A.1 Picture view of the experimental setup in the Lab (left side) and arrangement of the PV array (two modules) on the roof, free and shadowed by a white sheet (≅500 and 200 W/m^2, right side)

Table A.1 PV source parameters—standard test conditions (STC)

Parameter	Symbol	Value
Open circuit voltage	V_{OC}	43.4 V
Short circuit current	I_{SC}	4.8 A
Maximum photovoltaic voltage	V_{MPP}	34 V
Maximum photovoltaic current	I_{MPP}	4.4 A
Number of series-connected modules per string	N_S	2
Number of parallel strings	N_P	1

conditions (the sun irradiance on the PV module surface is ranging between 100 and 40% approx), as shown in the right side of Fig. A.1. The main parameters of the PV modules (Solar shell SQ 150-C) are given in Table A.1.

A.3 Power Converters

The power module used for the experimental tests is a three-phase Mitsubishi PS22A76 intelligent power IGBT dc/ac inverter. The top and bottom sides of the inverter are presented in Fig. A.2.

The main characteristics of the power module used are summarized in Table A.2.

The power board with the IGBT three-phase power module was available in the laboratory "SUN-Lab" at the department of Electrical, Electronic and Information

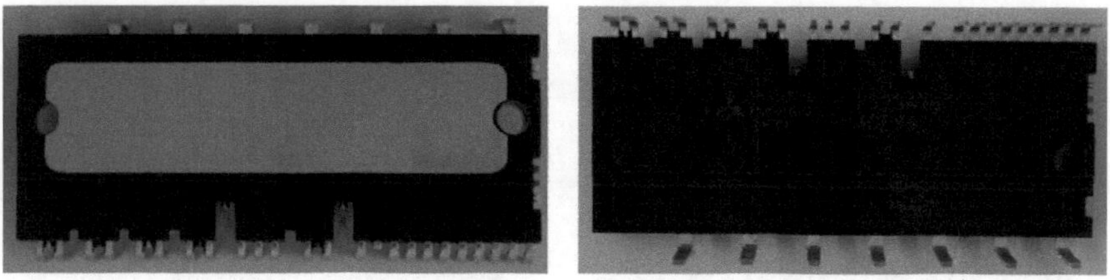

Fig. A.2 Mitsubishi PS22A76 three-phase inverter, (left) top side and (right) bottom side

Table A.2 Main characteristic of the power modules

Parameter	Value
Rated voltage	1200 V
Rated current	25 A
Maximum PWM input frequency	20 kHz

Engineering "Guglielmo Marconi", University of Bologna. Since LDN and the H-bridge structures are needed, only 1 or 2 legs have been used and the other leg(s) of the three-phase is grounded (it is not connected to the main circuit).

Figure A.3 depicts the top and the bottom sides of the power board. The dc-bus, the ac output (load) connection, control signal connections and an auxiliary control supply for the control part are all marked in the top side of the board. On the bottom side the power module is displayed. The power board has to be supplied by an isolated dc supply (auxiliary supply).

A.4 DSP Board

The Delfino TMS320F28377D has been used to generate the PWM signals of the inverter. F28377D is a powerful 32-bit floating-point microcontroller unit (MCU) designed for advanced closed-loop control applications such as industrial drives and servo motor control; solar inverters; digital power; transportation; and power line communications. The F28377D is a dual real-time control based on TI's 32-bit C28x floating-point CPUs, which provide 200 MHz of signal processing performance in each core. The DSP board is shown in Fig. A.4.

Table A.3 summarizes the main characteristics of DSP board F28377D used for the control part.

Code Composer Studio (CCS) and MATLAB/Simulink have been used to interact with the DSP board. In particular, Fig. A.5 shows the Simulink library of the DSP family F2837xD.

Fig. A.3 Top and bottom sides of the power board with a three-phase Mitsubishi PS22A76 intelligent power IGBT dc/ac inverter-power and control parts

Fig. A.4 Delfino TMS320F28377D

Table A.3 Main characteristic of DSP board (Delfino F28377D)

	F2837xD
C28x CPUs	2
Clock	200 MHz
Flash/RAM/OTP	512 Kw/102 Kw/2 Kw
On-chip Oscillators	✓
Watchdog Timer	✓
ADC	Four 12/16-bit (SOC)
Buffered DAC	3
Analog COMP/DAC	✓
FPU	✓ (each CPU)
6-Channel DMA	✓ (each CPU)
CLA	✓ (each CPU)
VCU/TMU	✓/✓ (each CPU)
ePWM/HRPWM	✓/✓
eCAP/HRCAP	✓/✓

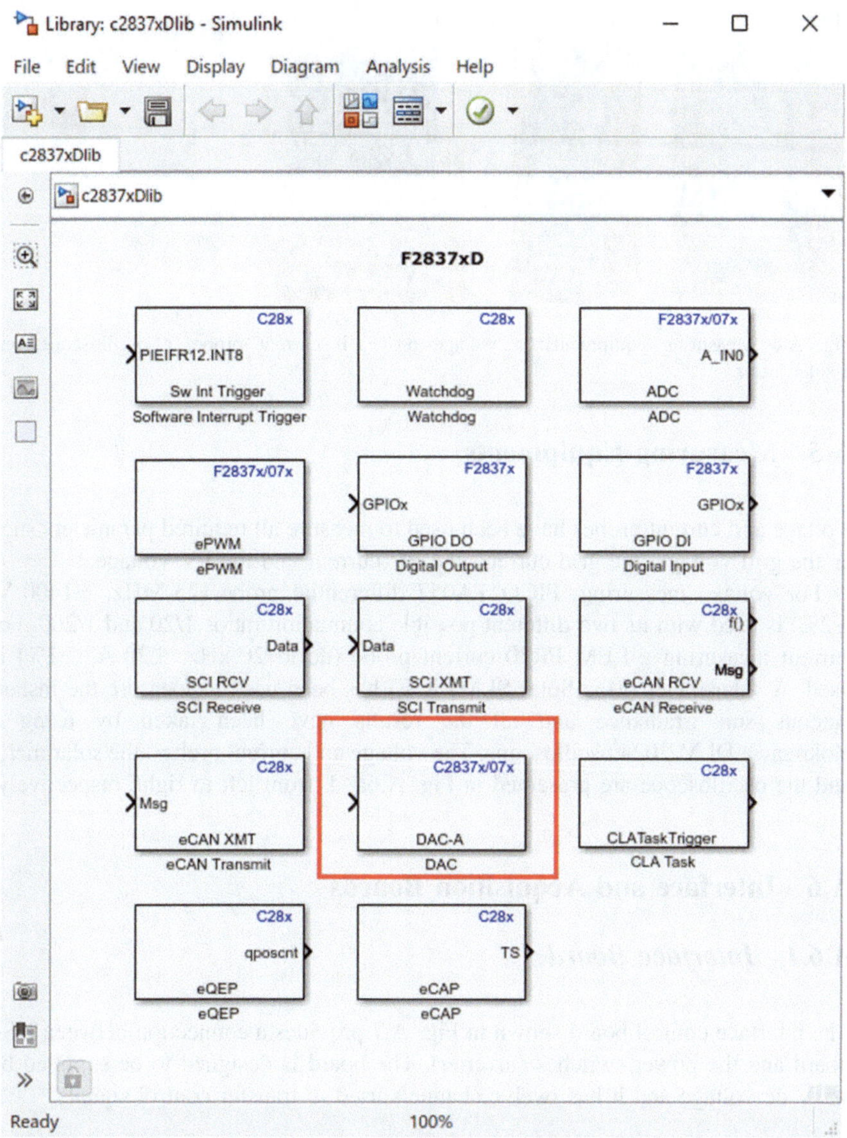

Fig. A.5 Simulink library of the DSP family TMS320F2837xD

The Simulink system model is compiled, converted into a C-code, and then automatically linked to the real-time TMS320F28377D processor through MATLAB/ Simulink. During the linking process, input–output (I/O) interface blocks, such as analog-to-digital converter (ADC), digital-to-analog converter (DAC) and the enhanced PWM (ePWM) should be included in the Simulink models. The behavior and performance of the inverter can be monitored in real time by using the CCS software.

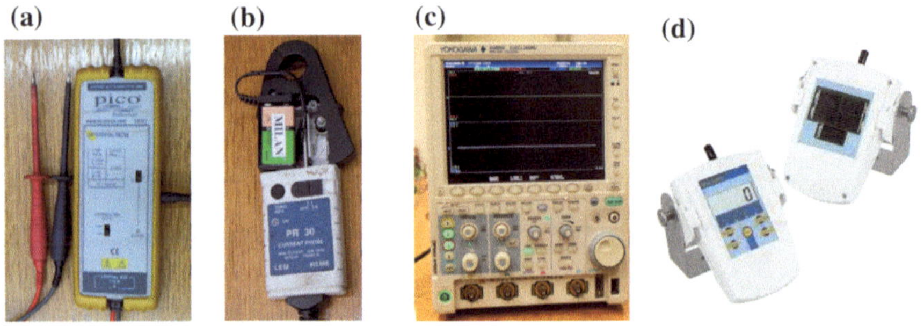

Fig. A.6 Measuring equipments: **a** voltage prove, **b** current probe, **c** oscilloscope, and **d** solar-meter

A.5 Measuring Equipments

Voltage and current probes have been used to measure all required parameters such us the grid voltage, the grid current, the PV current and the PV voltage.

For voltage measuring, PICO TA057 differential probe (25 MHz, ±1400 V, ±2%) is used with its two different possible attenuation ratios 1/20 and 1/200. For current measuring a LEM PR30 current probe (dc to 20 kHz, ±20 A, ±1%) is used. A solar-meter (Mac-Solar SLM18c-E) has been used to measure the instantaneous sun irradiance and all the results have been taken by using a Yokogawa DLM 2024 oscilloscope. The voltage and current probes, the solar meter and the oscilloscope are presented in Fig. A.6a–d, from left to right, respectively.

A.6 Interface and Acquisition Boards

A.6.1 Interface Board

The interface control board shown in Fig. A.7 provides a connection between DSP board and the power switches (inverter). The board is designed to be supplied by +5 V dc voltage and it has twelve channels used to transfer control signals.

A.6.2 Acquisition Board

Current and voltage transducers have been used for the measurement of voltage and current of both PV side and grid side inverter. Voltage and current sensors based on Hall effect are readily available. Two current transducers model LA 55-P (LEM® company) were used to measure PV and grid currents, while PV and grid voltages were measured using two voltage transducers model LV 25-P (LEM® company).

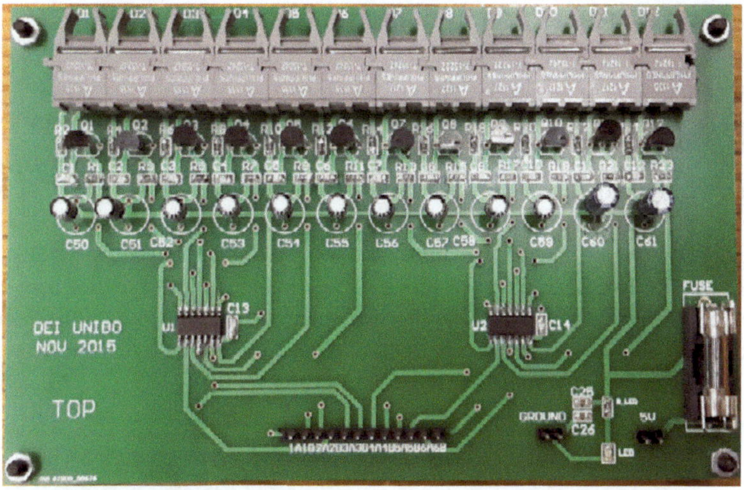

Fig. A.7 Interface board

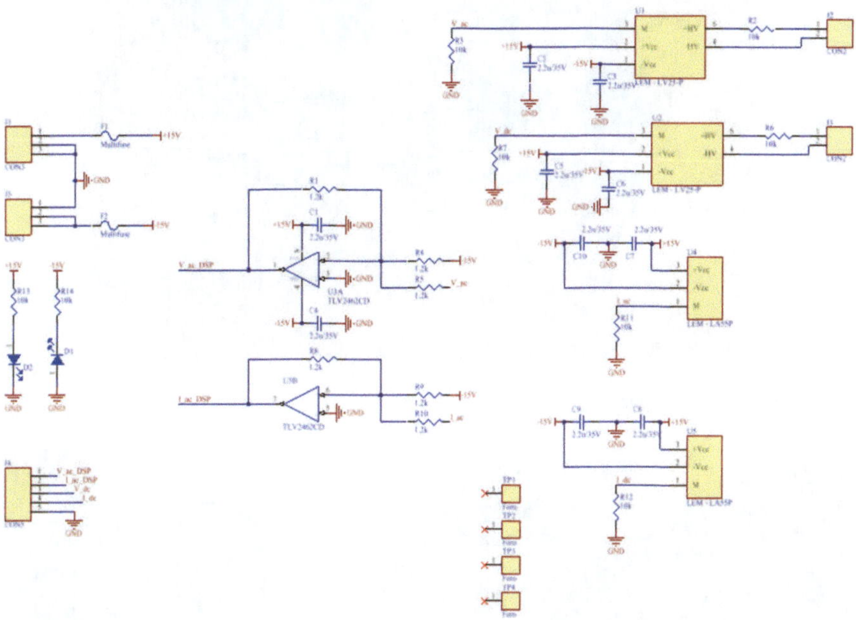

Fig. A.8 Altium scheme of the acquisition board (two voltage and two current transducers)

The acquisition system (Fig. A.9) has been designed by using the Altium software and a designed scheme with some details from the Altium program is given in Fig. A.8.

A.7 Transformers and Inductor for Grid Connection

The inverter has been connected to the grid (230 V, 50 Hz) by an isolation step-down transformer (230/48 V, 200 VA) and a Variac (0–240 V, 13.5 A), in order to adjust properly the voltage amplitude during the preliminary tests. The whole voltage ratio during the main tests was 10:1 approx (230/23 V). The ac-link inductor (8.8 mH) has been realized by a simple air-core coil, connected in series between the inverter and the isolation transformer, thought a safety bipolar power switch. The ac-link capacitor (25 uH) has been connected in parallel between the ac-link inductor and the isolation transformer. Figure A.10 shows these components. The three power resistors (20 Ω each) appearing on the left side of the picture have been used to implement the step changes on the dc side.

Fig. A.9 Acquisition boards

Fig. A.10 Grid connection: auto-transformer 0–240 V (Variac), isolation transformer 230/48 V (with bipolar switch), ac parallel capacitor, ac-link series inductor (air-core, on the floor)

Appendix B
Fourier Harmonic Analysis

B.1 Harmonic Analysis of Output Voltages

Due to the symmetry of the LDN output voltage, the normalized LDN output voltage averaged over the switching period, corresponding to the low-frequency components, can be written in terms of harmonics as

$$\bar{u}_{ac}^{L} \cong u_{ac}^{L} = U_0 + \sum_{k=2}^{\infty} u_k^L = U_0 + \sum_{k=2}^{\infty} U_k \cos(k\vartheta) \tag{B.1}$$

where U_0 is the average component and U_k is the amplitude of kth harmonic component, being k an even number.

The normalized modulating signal of the LDN cell can be written also as:

$$u_{ac}^{L} = \begin{cases} m|\sin(\vartheta)|, & m|\sin(\vartheta)| \leq 0.5 \\ 1 - m|\sin(\vartheta)|, & m|\sin(\vartheta)| \geq 0.5 \end{cases} \tag{B.2}$$

In case of $m|\sin\vartheta| \leq 0.5$, the average component U_0 is given as:

$$U_0 = \frac{2m}{\pi} \tag{B.3}$$

In case of $m|\sin\vartheta| \geq 0.5$, the normalized LDN voltage can be written as follows:

$$u_{ac}^{L} = \begin{cases} m\sin\vartheta, & 0 \leq \vartheta \leq \arcsin(1/2m) \\ 1 - m\sin\vartheta, & \arcsin(1/2m) \leq \vartheta \leq \frac{\pi}{2} \end{cases} \tag{B.4}$$

M. Hammami, *Level Doubling Network and Ripple Correlation Control MPPT Algorithm for Grid-Connected Photovoltaic Systems*, Springer Theses,
https://doi.org/10.1007/978-3-030-10492-4

Integrating (B.4) over π/2, the average component can be calculated as:

$$U_0 = \frac{2}{\pi}\left[\int_0^{\alpha_m} m \sin\vartheta \, d\vartheta + \int_{\alpha_m}^{\pi/2} (1 - m \sin\vartheta) d\vartheta\right] \tag{B.5}$$

being $\alpha_m = \arcsin(1/2m)$.
Leads to:

$$U_0 = \frac{2}{\pi}\left[m + \frac{\pi}{2} - \sqrt{4m^2 - 1} - \arcsin\left(\frac{1}{2m}\right)\right] \tag{B.6}$$

As a result, the average component of the normalized LDN output voltage can be expressed as:

$$U_0 = \begin{cases} \frac{2}{\pi} m, & m \le 0.5 \\ \frac{2}{\pi}\left[m + \frac{\pi}{2} - \sqrt{4m^2 - 1} - \arcsin\left(\frac{1}{2m}\right)\right], & m \ge 0.5 \end{cases} \tag{B.7}$$

Knowing that (B.2) is an even function [only cosine terms as expressed in (B.1)], the amplitudes of the low frequency components can be calculated as:

$$U_k = \frac{2}{T}\int_0^T u_{ac}^L \cos\left(\frac{k2\pi t}{T}\right) dt = \frac{1}{\pi}\int_0^{2\pi} u_{ac}^L \cos(k\vartheta) d\vartheta \tag{B.8}$$

In case of $m\,|\sin\vartheta| \le 0.5$, the amplitude of the kth harmonic component U_k of (B.2) can be expressed as:

$$U_k = -\frac{4m}{\pi(k^2 - 1)} \quad k \text{ even} \tag{B.9}$$

In case of $m|\sin\vartheta| \ge 0.5$, U_k has been calculated as:

$$U_k = \frac{4}{\pi}\int_0^{\pi/2} u_{ac}^L \cos(k\vartheta) d\vartheta \tag{B.10}$$

$$U_k = \frac{4}{\pi}\left[\int_0^{\alpha_m} m \sin\vartheta \cos(k\vartheta) d\vartheta + \int_{\alpha_m}^{\pi/2} (1 - m \sin\vartheta) \cos(k\vartheta) d\vartheta\right] \tag{B.11}$$

Solving (B.11) leads to:

$$U_k = \frac{4}{\pi}[A_k + B_k] \tag{B.12}$$

The coefficients A_k and B_k can be calculated as:

$$A_k = \int_0^{\alpha_m} m \sin\vartheta \cos(k\vartheta) d\vartheta \tag{B.13}$$

$$A_k = \frac{1}{2k^2 - 2}\left[k \sin(k\alpha_m) + \sqrt{4m^2 - 1}\cos(k\alpha_m) - 2m\right] \tag{B.14}$$

$$B_k = \int_{\alpha_m}^{\pi/2} (1 - m \sin\vartheta) \cos(k\vartheta) d\vartheta \tag{B.15}$$

$$B_k = -\frac{1}{k}\sin(k\,\alpha_m) + \frac{1}{2k^2 - 2}\left[k \sin(k\,\alpha_m) + \sqrt{4m^2 - 1}\cos(k\,\alpha_m)\right] \tag{B.16}$$

To conclude, the amplitudes of the U_k can be expressed as:

$$U_k = \begin{cases} -\frac{4m}{\pi(k^2-1)}, & m \leq 0.5 \\ \frac{4}{\pi}[A_k + B_k], & m \geq 0.5 \end{cases} \quad k \text{ is even} \tag{B.17}$$

B.2 Harmonic Analysis of Input Currents

For the H-bridge, the modulating signal can be obtained as

$$\bar{u}_{ac}^H \cong u_{ac}^H = u_{ac} - u_{ac}^L \tag{B.18}$$

Both H-bridge and LDN dc-link voltages can be considered almost constant. Being V and $V/2$ the corresponding average components (over the fundamental period), the normalized H-bridge and LDN voltages become 1 and 1/2, respectively. With this assumption, and considering the averaged quantities over the switching period, the power balance for LDN and H-bridge cells can be written as:

$$\begin{cases} \frac{1}{2}\bar{i}^L = u_{ac}^L\,\bar{i}_{ac} \\ \bar{i}^H = u_{ac}^H\,\bar{i}_{ac} \end{cases} \tag{B.19}$$

being $\bar{i}_{ac}$ the instantaneous output current expressed as:

$$i_{ac} = I_{ac}\ \sin\vartheta \approx \bar{i}_{ac} \tag{B.20}$$

Replacing (B.1), (B.18) and (B.20) in (B.19), the harmonic spectra of LDN and H-bridge input currents are calculated as:

$$\bar{i}^{L} = \left[(2U_0 - U_2)\sin\vartheta - \sum_{n=3}^{\infty}(U_{n+1} - U_{n-1})\ \sin n\vartheta\right] I_{ac} \tag{B.21}$$

$$\bar{i}^{H} = \frac{1}{2}\left[m + (2U_0 - U_2)\ \sin\vartheta - m\ \cos 2\vartheta - \sum_{n=3}^{\infty}(U_{n+1} - U_{n-1})\ \sin n\vartheta\right] I_{ac} \tag{B.22}$$

From Eq. (B.21), the amplitudes of the low-frequency input LDN current harmonic components are:

$$\begin{cases} I^{L} = I_2^{L} = 0 \\ I_1^{L} = (2U_0 - U_2)\, I_{ac} \\ I_n^{L} = |U_{n+1} - U_{n-1}|\, I_{ac} \end{cases} \quad n \geq 3 \text{ odd integer} \tag{B.23}$$

From Eq. (B.22), the amplitudes of the low-frequency input H-bridge current harmonic components are:

$$\begin{cases} I^{H} = I_2^{H} = \dfrac{1}{2} m\, I_{ac} \\ I_1^{H} = \dfrac{1}{2}(2U_0 - U_2)\, I_{ac} \\ I_n^{H} = \dfrac{1}{2}|U_{n+1} - U_{n-1}|\, I_{ac} \end{cases} \quad n \geq 3 \text{ odd integer} \tag{B.24}$$

Appendix C
Extension to General Power Factor Angle

The calculation of dc-link current and voltage harmonics for H-bridge cell can be extended from unity power factor to the case of general power factor angle as follows:

C.1 Low-Frequency (Averaged) Current Component for H-Bridge Cell

The analytical developments of Chap. 4 have been extended to an output current with a general load phase angle. In this case, the instantaneous output current is:

$$i_{ac} = I_{ac}\ \sin(\vartheta - \varphi) \approx \bar{i}_{ac} \tag{C.1}$$

The averaged input H-bridge current is calculated based on (4.4), (4.5), (C.1) and (4.3), as:

$$\bar{i}^H = \tilde{i}^H + I^H = \begin{cases} I_{ac}\ \sin(\vartheta - \varphi)\ (m\ \sin\vartheta - m\,|\sin\vartheta|), & m\,|\sin\vartheta| \le 0.5 \\ I_{ac}\ \sin(\vartheta - \varphi)\ (m\ \sin\vartheta - 1 + m\,|\sin\vartheta|), & m\,|\sin\vartheta| \ge 0.5 \end{cases} \tag{C.2}$$

M. Hammami, *Level Doubling Network and Ripple Correlation Control MPPT Algorithm for Grid-Connected Photovoltaic Systems*, Springer Theses,
https://doi.org/10.1007/978-3-030-10492-4

The H-bridge input current can be calculated in terms of harmonics, respectively, as:

$$\bar{i}^H = \frac{1}{2} I_{ac} \left\{ \begin{array}{l} m \cos\varphi + [U_2 \sin(\vartheta + \varphi) - 2\,U_0 \sin(\vartheta - \varphi)] - m \cos(2\vartheta - \varphi) \\ + \sum_{n=3}^{\infty} [U_{n+1} \sin(n\vartheta + \varphi) - U_{n-1} \sin(n\vartheta - \varphi)] \end{array} \right\} \tag{C.3}$$

where $n \geq 3$ is odd, and the terms U_0, U_2 and U_n are defined in the Appendix B.

From Eq. (C.3), the amplitudes of the low-frequency input H-bridge current harmonic components in case of general phase angle are:

$$\begin{cases} I^H = \frac{1}{2} I_{ac} m \cos\varphi \\ I_1^H = \frac{1}{2} I_{ac} \sqrt{(\cos\varphi)^2 (U_2 - 2U_0)^2 + (\sin\varphi)^2 (U_2 + 2U_0)^2} \\ I_2^H = \frac{1}{2} I_{ac} m \\ I_n^H = \frac{1}{2} I_{ac} \sqrt{(\cos\varphi)^2 (U_{n+1} - U_{n-1})^2 + (\sin\varphi)^2 (U_{n+1} + U_{n-1})^2} \end{cases} \tag{C.4}$$

It can be noted that the 2nd order harmonic is constant (only depends on the modulation index), where the 1st is changing with the output phase angle.

Figure C.1 shows the first harmonic amplitudes of H-bridge input current as a function of modulation index and output phase angle.

It can be noted that near the working conditions (the corresponding modulation index is equal to m = 0.82), the amplitudes of first harmonic H-bridge input current is not a function of the output phase angle φ.

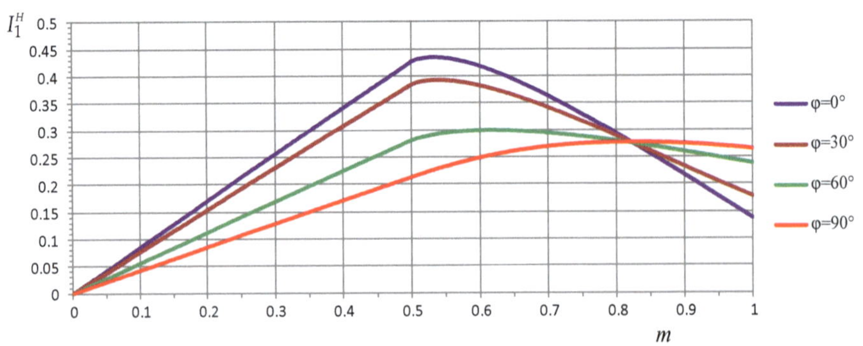

Fig. C.1 First harmonic amplitudes of H-bridge input current for sinusoidal unity output current as a function of the modulation index and output phase angle φ = 0°, 30°, 60° and 90°

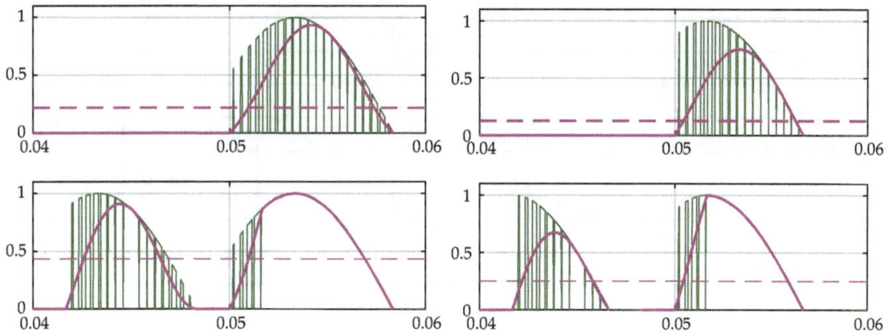

Fig. C.2 Input current and its averaged counterpart for $m = 0.5$ (top) and $m = 1$ (bottom) of the LDN cell (left) and H-bridge cell (right) in case of unity output current ($I_{ac} = 1$ A) and output phase angle $\varphi = 30°$ (left) and $60°$ (right)

The instantaneous input current, its low-frequency component corresponding to the averaged value over the switching period, and its dc component are shown in Fig. C.2 for H-bridge cell for two cases of output phase angle.

The H-bridge low-frequency components shown in Fig. C.2 are calculated by Eq. (C.2).

C.2 Low-Frequency Voltage Component for H-Bridge Cell

The amplitudes of the low-frequency input voltage components at kth harmonic V_k^H can be calculated on the basis of the amplitudes of the low-frequency input current components (C.4), and the dc-link equivalent impedance as:

$$V_k^H = Z_k^H \, I_k^H \tag{C.5}$$

being Z_k^H the parallel between the reactance of the dc-link capacitor C_H, and the equivalent input resistance R_s, calculated at the kth harmonic order:

$$Z_k^H = \frac{R_s}{\sqrt{k^2 R_s^2 \omega^2 C_H^2 + 1}} \tag{C.6}$$

Considering realistic parameters of a photovoltaic module, the equivalent series resistance R_s is always larger than $(1/k\omega C_H)$. In this case, Eq. (C.6) becomes:

$$Z_k^H = \frac{1}{k\omega C_H} \tag{C.7}$$

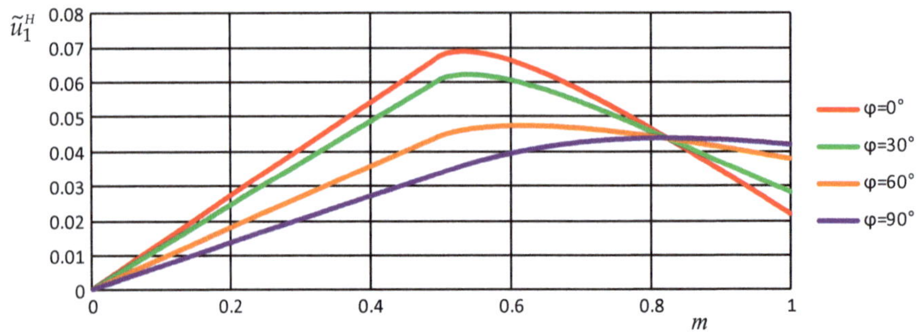

Fig. C.3 Normalized first harmonic amplitude of H-bridge input voltage for unity output current as a function of the modulation index and output phase angle φ = 0°, 30°, 60° and 90°

Introducing (C.7) in (C.5) and considering (C.4), the amplitudes of the low-frequency input voltage components at the kth harmonic become:

$$V_k^H = \frac{1}{k\omega C_H} I_k^H \tag{C.8}$$

Considering only the first and second harmonic order, the voltage amplitudes can be expressed as:

$$V_1^H = \frac{1}{\omega\, C_H} \frac{1}{2} I_{ac} \sqrt{(\cos\,\varphi)^2 (U_2 - 2U_0)^2 + (\sin\,\varphi)^2 (U_2 + 2U_0)^2} \tag{C.9}$$

$$V_2^H = \frac{1}{2\omega C_H} \frac{m}{2} I_{ac} \tag{C.10}$$

The first harmonic amplitude of H-bridge input voltage can be calculated on the basis of (C.9) as:

$$V_1^H = \frac{I_{ac}}{fC_H} U_1^H \tag{C.11}$$

being U_1^H the normalized first harmonic amplitude of H-bridge input voltage expressed as:

$$U_1^H = \frac{1}{4\pi} \sqrt{(\cos\varphi)^2 (U_2 - 2U_0)^2 + (\sin\varphi)^2 (U_2 + 2U_0)^2} \tag{C.12}$$

The normalized first harmonic amplitude of H-bridge input voltage is represented in Fig. C.3 as a function of the modulation index, for different output phase angles φ.

It can be noticed that, when the following condition occurs

$$U_2 = 0 \tag{C.13}$$

the normalized first harmonic amplitudes of H-bridge input voltage is not a function of the output phase angle φ. By solving (C.13) the corresponding modulation index is equal to $m = 0.82$, and correspondingly

$$U_1^H(m = 0.82) \cong 0.043. \tag{C.14}$$

Author Biography

Dr. Manel Hammami is currently a Postdoctoral Fellow at the Department of Electrical, Electronic and Information Engineering, "Guglielmo Marconi"—DEI, University of Bologna, in Prof. Gabriele Grandi's Power Electronics team, working on the development of high-power converters for fast charging of electric vehicles. She received the Bachelor (2010) and Master degree (2012), both in Electrical Engineering, from the Faculty of Engineering, Department of Electrical Power Systems, University of Tunis, Tunisia. In 2014, She enrolled in the Ph.D. program in Electronics, Telecommunications and Information Technologies Engineering of DEI, under the supervision of Profs. Gabriele Grandi and Massimo Rudan. She was supported by the Erasmus Project "EU-Metalic II". In April 2018, she defended a Ph.D. thesis entitled "Level Doubling Network and Ripple Correlation Control MPPT Algorithm for Grid-Connected Photovoltaic Systems", receiving the highest rate "Excellent" from the evaluation board; this book is the follow-up of the Ph.D. thesis.

Manel Hammami authored a number of papers published in high-ranking Journals or presented at IEEE conferences; she is an appreciated reviewer for three reputed international journals (IEEE Trans. on Industrial Electronics, IET Power Electronics, and MDPI Energies), being recognized as a top peer reviewer for 2017 in the Publons Peer Review Awards (6th in the ranking of the whole University of

M. Hammami, *Level Doubling Network and Ripple Correlation Control MPPT Algorithm for Grid-Connected Photovoltaic Systems*, Springer Theses,
https://doi.org/10.1007/978-3-030-10492-4

Bologna). Her research interests include photovoltaic generation systems, multilevel converters and on-board charging stations for electric vehicles.

Manel Hammami's university profile and publications are available at https://www.unibo.it/sitoweb/manel.hammami2/.